Carbon Management in the Built Environment

T0221407

Three broad sectors of the economy are generally recognised as key to a low carbon future: energy, construction and transportation. Of these, carbon management in the built environment remains the least well studied.

This much-needed book brings together the latest developments in the field of climate change science, building design, materials science, energy and policy in a form readily accessible to both students of the built environment and practitioners. Although several books exist in the broad area of carbon management, this is the first to bring together carbon management technology, technique and policy as they apply to the building sector.

Clear and succinct sections on the overarching principles, policies, approaches and technologies are combined with case studies and more in-depth coverage of the most relevant topics. It explains how to produce a simple carbon footprint calculation, while also being an informative guide for those developing or implementing more advanced approaches. This easy to read book is the ideal primer for anyone needing to get to grips with carbon management in the built environment.

Rohinton Emmanuel is a Reader in Sustainable Design and Construction and the Director of the Centre for Energy and the Built Environment at Glasgow Caledonian University, UK. He has pioneered the inquiry of urban climate change in warm regions and has taught and consulted on climate and environment sensitive design, building energy efficiency, thermal comfort. He has authored over 50 research papers in the areas of climate change in the built environment, building and urban energy efficiency and thermal comfort, and a book related to these efforts, *An Urban Approach to Climate Sensitive Design: Strategies for the Tropics*, was published by Routledge in 2005.

Keith Baker is a Researcher in Sustainable Urban Environments at the Centre for Energy and the Built Environment at Glasgow Caledonian University, UK. He is also a member of the Scottish Carbon Accounting Group, and currently directs Sustainable Footprints, a carbon and sustainability management consultancy. Keith's main research interests are in climate change, energy, the built environment, and the environmental impact of technology.

Carbon Management in the Built Environment

Rohinton Emmanuel and Keith Baker

Routledge
Taylor & Francis Group

LONDON AND NEW YORK

First published 2012
by Routledge
4 Park Square, Milton Park, Abingdon, Oxon OX14 4RN
605 Third Avenue, New York, NY 10017

Routledge is an imprint of the Taylor & Francis Group, an informa business

British Library Cataloguing in Publication Data
A catalogue record for this book is available from the British Library

Library of Congress Cataloging in Publication Data
  Emmanuel, Rohinton.
  Carbon management in the built environment /
  Rohinton Emmanuel, Keith Baker.
    p. cm.
  Includes bibliographical references and index.
  1. Sustainable buildings. 2. Sustainable construction.
  I. Baker, Keith, 1979– II. Title.
  TH880.E485 2012
  690.028'6   dc23
  2011050648

ISBN: 978–0–415–68406–4 (hbk)
ISBN: 978–0–415–68407–1 (pbk)
ISBN: 978–0–203–80331–8 (ebk)

Typeset in Garamond
by Swales & Willis Ltd, Exeter, Devon

To Melanie and Cindy
For your many sacrifices that made our task light

Contents

Figures

Tables

Foreword

This is a very welcome book indeed: a clear, comprehensible and sensible book on the complex subject of carbon accounting and management in the built environment will help a generation of building designers and managers to come to grips with how, in reality, to reduce energy emissions from both new and existing buildings. The oft repeated maxim that we have all seen many low carbon building designs, but very few really low carbon buildings, does hold true. This book, which explains the key carbon issues and methods as applied to the built environment, is important in informing the way we move on from the high energy, high carbon, twentieth-century business-as-usual approaches to urban design and towards the low carbon buildings so essential for a safer future in the difficult decades ahead.

I believe that the division of the design professions in the twentieth century into silos of responsibility has had devastating effects on the quality and performance of the buildings we produce. Many of the generation of young architects now leave universities around the world being unable to design basic buildings, let alone low carbon ones. A generation of building services engineers were not even taught how to shape and design low energy, passive buildings that could be naturally ventilated, in the push to get more and more servicing 'product' into buildings. In the UK and elsewhere the entire building regulation system is riddled with perverse incentives that discourage truly low carbon design – as evidenced by the UK building regulations that enable a simple office design to pass their standards while the same naturally ventilated office on a green-field site will fail. The results of such wrongly facing trends can be seen on the high streets of cities around the world, and a growing number of buildings built two, three or four decades ago are already being pulled down today, or lie empty awaiting some uncertain fate.

This book is part of the new, universal language of low carbon building design and management. It is an important book for that. It explains why the transition to low carbon built environments is so important, the international landscape of action, and the regulation that drives the groundswell of change in our industry that affects us all. Whether you are a climate change sceptic or believer, you now have to know about how to count and account for carbon, and how to design and manage buildings to reduce carbon emissions from them. This book tells you how.

Fundamental to how we design our buildings and their systems to reduce energy and related emissions is *why* we do so. A vital strand to the language of low carbon buildings is a re-evaluation of comfort and how to achieve it in low carbon buildings and the understanding of comfort as a goal for good design as opposed to a product produced by machines. Another strand is the inclusion of a wide range of new issues into the way we account for the carbon of a development, including transport, waste and many other constituent elements of a development that must now be accounted for in assessing the carbon impacts of buildings.

That is why I am so pleased that this book has been written to provide a first rate and eminently usable guide to carbon accounting and management. Carbon accounting and management, used to baseline and benchmark emissions in a process of continual performance improvement, provide the glue that sticks together the disparate issues in the final building account. Trust me – the issues are complex, as we have found out in ICARB, the Initiative for Carbon Accounting, in which the authors of this book have also played a major part.

But it is not only students on specialist low carbon building design and management courses who will be able to use this as their course textbook of choice, but undergraduates and all design professionals, whether architects or engineers, because carbon management is a core strand of the new language of low carbon buildings, and we all need to learn and share that language.

Susan Roaf
Professor of Architecture, Heriot-Watt University,
and founder of the Initiative for Carbon Accounting (ICARB)
December 2011

Preface

This handbook is our first attempt to provide an overview of the many issues around managing carbon and greenhouse gases in the built environment, and we hope it will be useful as a primer for anyone new to the field and a reference guide for students and professionals alike. The hardest part of writing this book has been trying to strike the best balance between the scope and the depth of the coverage, and because of this some of the final contents have changed somewhat from our initial ideas. We hope these changes have been for the better in enabling a more specific coverage of the issues most relevant to the built environment, and we welcome any feedback for future editions.

The opening section of the book summarises the contexts in which those working to reduce emissions from the built environment operate – from global governance and climate change down to the more practical issues around reducing emissions from the built environment. The second section presents a more detailed coverage of these latter issues and how they may be addressed in specific built environmental contexts (new build, existing build and cities). By opening this section with energy generation we underline the ultimate dependence of all emissions reduction targets on decarbonising our energy supplies. The final section covers many of the protocols, standards, approaches, methods, tools and techniques that can, or must, be adhered to or employed as part of assessing energy consumption in the built environment and the emissions attributable to it. Identifying and selecting the most relevant of these have been a daunting task, and we apologise for any omissions and being mainly limited to English language sources.

Although we have included some illustrative case studies, it was never our intention to provide a template example for use in carbon and greenhouse gas accounting. In reality carbon and GHG assessments (or 'footprints') vary widely according to factors such as their subjects, aims, methods, tools, results and outputs, as well as any legislative requirements they are subject to – and any template would rapidly become out of date. Similarly, whilst there are many perfectly good commercial carbon accounting tools and services on the market we have not attempted to summarise or recommend any of these; not only will different projects require different tools, but also different users will judge those available on different criteria. However, we hope this handbook contains sufficient guidance and pointers to key publications to enable the development of carbon and GHG assessments for most common aspects of the built environment, whether users decide to use existing tools or develop their own.

Carbon and GHG management is a rapidly evolving, complex and contested field, and we are conscious that by even writing about some issues we are opening ourselves to accusations of bias. The issue of energy generation provides a case in point. The options for the scope of what to include in this chapter ranged from limiting it to building-integrated technologies up to a full coverage of all existing and possible future generation technologies. As with carbon and GHG accounting, the key problem was deciding where to set the boundaries. We have avoided discussing future technologies because of the uncertainties around them, and also because the urgent need to tackle climate change means that we should not let predictions of the potential of future technologies cloud our judgements over decisions that must be based on what works today. We could have also dodged the bullet of nuclear power by limiting our scope in a number of ways, but all of these would have required omitting other technologies which are also contributing to reducing emissions from the built environment.

A more building-specific issue that can provoke heated debate is providing heating, ventilation and air conditioning (HVAC). It is difficult to work in this field and not become an advocate of one or more approaches to meeting these demands. However, regardless of whether you favour passive or mechanical ventilation, or high thermal mass or light build, the priority should be identifying the most effective solutions in any given context. The same applies to identifying the most appropriate options for retrofitting buildings for energy efficiency: there is no one-size-fits-all approach, but some approaches are more generally applicable than others, and conversely some buildings are more individual than others. Identifying and implementing those solutions require pulling together the best available evidence, and we hope that this book will be a useful aid for doing so.

In 2004 the GHG Protocols Group set out five key principles of carbon accounting: relevance, consistency, completeness, transparency and accuracy. So how well do we think this text compares against them? With so much information that could be captured we'd never claim it was 100 per cent complete, and its accuracy will decline as the latest information changes over time, but we hope readers find it relevant and consistent, and we have done our utmost to ensure that it's transparent. We hope you'll find it useful, and we leave you with a little bit of satire for when the going gets tough and the figures simply refuse to add up.

R. Emmanuel and K. Baker
Glasgow, December 2011

A Carbon Accountant's Completely Perfect and Absolutely Quantitative Method of Measuring His Emissions
By Keith Baker – with apologies to R. Landon and A. Bueche

I change our energy bills to carbon with factors I can't deduce,
Differentiate by consumption, determine frequency of use;
Where uncertainty arises I simply calculate the mean,
(And then deduct a small percentage for electricity that's green).

I integrate our recycling rate upon a monthly basis;
Calculate just what our place in the race to zero waste is;
And our emissions inventory has boundary conditions,
Whose final calibration is the Company's net emissions.

And thus I create numbers where there were none before;
I have lots of facts and figures – and formulae galore –
And these quantitative studies make the whole thing crystal clear,
Our emissions will be exactly 10 per cent lower than last year.

Acknowledgements

A comprehensive text on carbon management specifically focused on the built environment does not exist at present. Our efforts therefore relied on many friends and colleagues for intellectual stimulation, sources of information, and discussion and debate. We are particularly indebted to Tony Kilpatrick, Head, Department of Construction and Surveying, School of Engineering and Built Environment, Glasgow Caledonian University, for allowing us the time to write the book, and to our colleagues: Drs Paul Baker, Ole Pahl, Mark Phillipson, Paul Teedon and Craig Thomson. We are honoured to have Professor Sue Roaf, Heriot-Watt University, Edinburgh, writing the Foreword and, together with Sam Chapman and Richard Roaf, for providing encouragement and advice via the Initiative for Carbon (ICARB), Edinburgh. Dr Liz Hawkins, Community Analytical Services, Scottish Government, directed us to many useful sources in Scotland. All help provided by Gary Davis, Director, Ecometrica, Edinburgh and Montreal, Scott Herbert, University of Leicester, England, and Professor Katherine Irvine and Professor Mark Rylatt, De Montfort University, Leicester, England, is gratefully acknowledged.

We are indebted to the following for kind permission to reproduce illustrations used in the book: Figure 1.1 (Intergovernmental Panel on Climate Change, Working Group III, World Meteorological Organization; Ms Sophie Schingemann); Figure 2.2 (World Resources Institute, Washington, DC, USA, Office of Science and Research: Ms Ashleigh Rich); Figure 2.3 (Global Commons Initiative: Aubrey Meyer); Figure 2.4 (Professor Aviel Verbruggen, University of Antwerp, Belgium); Figure 2.5 (UNEP/GRID-Arendal); Figure 2.6 (Global Carbon Project: Dr Josep Canadell); Figure 3.1 (Scottish Government, Soils and Contaminated Land Team, Environmental Quality Division: Mr Francis Brewis); Figure 3.3 (Department for Energy and Climate Change: Ms Lisa Ahmad; and Department for Environment, Food and Rural Affairs, Press Office, Communications Group: Mr Paul Leat); Figure 5.2 (University of Veterinary Medicine Vienna, Biometeorology and Mathematical Epidemiology Group, Institute for Veterinary Public Health: Professor Franz Rubel); Figure 6.1 (University of Moratuwa, Sri Lanka, Department of Civil Engineering: Professor Thishan Jayasinghe); Figure 10.2 (Scottish Government, Scottish Buildings Standards Agency: Dr Fraser Walsh). The following gave us permission to reproduce photographs, for which we are very grateful: Mr Chris Morgan, Locate Architects, Dunblane, Scotland (Figure 5.7 and cover photo); Mr Vidhura Ralapanawe, MAS Thurulie Ltd, Sri Lanka (Figure 5.8); Dr Hans Rosenlund, CEC Design, Sweden (Figures 5.9 and 5.10); Ms Veronica Olivotto, Institute for Housing and Urban Development Studies (IHS), Erasmus University, Rotterdam, and Dr Denis Fan, Faculty of Technology, De Montfort University, Leicester, England (Figure 8.2).

Section A

Overview

1 Historical background

From sustainable development to carbon management

While the phenomenon of global climate change is largely responsible for the current focus on carbon management, one must be mindful of the wider implications of carbon emission to sustainable development and the role the built environment plays in this interaction. In spite of all the attention on carbon management in recent years, the fact remains that global greenhouse gas emissions and the global carbon intensity (measured as carbon emission per unit of economic output) have continued to rise (Pielke, 2010). The world emitted twice as much carbon dioxide per marginal unit of economic activity in the decade leading to 2008 than in the previous decade (Prins *et al.*, 2009). It seems that global economic output is unable to extricate itself from carbon dependency (99 per cent of the variations in carbon emissions can be explained by the changes in the approximately USD 50 trillion global economy – Pielke, 2010), and the trend is unlikely to reverse. This is made clear by the 'Kaya Identity' (Kaya, 1990), which is composed of two primary factors: economic growth and technology changes.

1   Carbon dioxide emissions = Population × Per capita GDP × Energy intensity × Carbon intensity
2   P = Total population
3   GDP/P = Per capita GDP
     GDP = Economic growth (contraction)
           = P × GDP/P = GDP
4   Energy intensity (EI) = TE/GDP = Total energy (TE) consumption/GDP
     Carbon intensity (CI) = C/TE = Carbon emissions/total energy consumption
5   EI × CI = 'Carbon intensity of the economy' = TE/GDP × C/TE = C/GDP

Thus, according to the logic of these relationships, carbon accumulating in the atmosphere can be reduced only by reducing one or more of the following: population, per capita GDP, energy intensity or the carbon intensity of the economy.

It is at this point that the wider importance of 'sustainable development' comes into play. The definition of 'sustainable development' is by now well known: the Brundtland Commission

Report (WCED, 1987) defines sustainable development as 'development that meets the needs of the present without compromising the ability of future generations to meet their own needs'. It contains within it two key concepts:

1. the concept of 'needs', in particular the essential needs of the world's poor, to which overriding priority should be given;
2. the idea of limitations imposed by the state of technology and social organisation on the environment's ability to meet present and future needs (WCED, 1987).

The built environment is critical to both of these concepts of 'sustainable development', and therefore the management of carbon in the built environment is central to our efforts to bequeath a 'sustainable' world to future generations. Buildings (especially housing, but also other infrastructure) contribute to fulfilling the 'need' for sustainable development, especially for the poor; the state of technology in the built environment provides a quick win for the world to achieve a low carbon (and therefore sustainable) future. Hence this book.

1.1 The built environment's role in the global carbon cycle

The built environment is a major consumer of energy and thus a significant contributor of greenhouse gases (GHGs). The United Nations estimates that buildings consume 30–40 per cent of the total energy used worldwide (UNEP, 2007). If we include cities, up to 90 per cent of energy use occurs in and/or for cities (Svirejeva-Hopkins, Schellnhuber and Pomaz, 2004). Given the rapid urbanisation and associated development in built infrastructure in developing nations, the role of the built environment in energy use and therefore GHG emissions is likely to be dramatic (Figure 1.1). This is especially the case in Asia, but also in Latin America and, to a lesser extent, in sub-Saharan Africa. At the same time, the technical know-how needed to achieve substantial savings in energy use in the built environment (and therefore large reductions in GHG emissions) is largely well known. Therefore, in theory at least, the expected boom in the built infrastructure in the developing world could potentially offer huge opportunities to reduce emissions and wean the world away from its carbon intensive ways. What's more, the IPCC's AR4 estimated that doing so in the built environment sector is not only the most technically feasible but also cost-effective – in fact AR4 asserts that the lifecycle costs are negative (Levine *et al.*, 2007); 5.0 Gt of CO_2e of the total likely savings of 6.10 Gt of CO_2e at a negative cost (i.e. cost savings) is estimated to come from the buildings sector (see Table 1.1). This represents over 20 per cent of all the CO_2e estimated technically feasible to save at the global scale.

What then is preventing such huge and cost-effective potential from being realised? This book is an attempt to present the case for emission reduction in the built environment and examine the barriers and opportunities for their implementation. It presents the policy, regulatory and best-practice landscape of emission reduction in the built environment in the context of real-world challenges for achieving them. Given the realities of energy use in the built environment (approximately 80–90 per cent in the lifetime operation and maintenance plus 10–20 per cent embodied in the manufacturing, construction and demolition of the built assets – see UNEP, 2007), the book presents the case for operational energy efficiency in buildings both new and existing as well as the built environmental context in cities. It also presents the assessment regimes, protocols and regulatory frameworks for emission management in the built environment by way of case studies and exemplars. These occur in the context of delineating the key technologies for carbon management in key economic sectors and are rounded off with likely new and emerging technologies with the greatest scope for further reduction. The social and

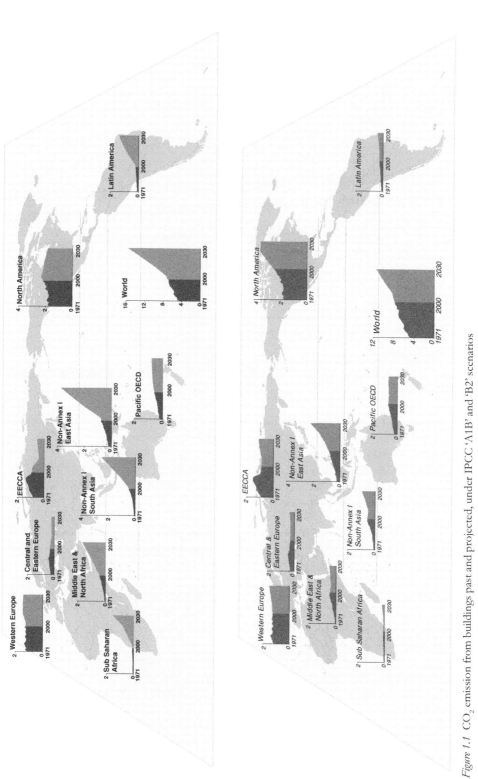

Figure 1.1 CO$_2$ emission from buildings past and projected, under IPCC 'A1B' and 'B2' scenarios

Source: Levine *et al*, 2007.

Table 1.1 Costs of GHG mitigation for different economic sectors

Sector	Region	Cost of potential mitigation (in USD / tCO_2e)			
		< 0	0–20	20–50	50–100
		Gt CO_2e			
Energy supply	OECD		0.90	0.50	0.00
	World		1.90	1.40	0.35
Transport	OECD	0.25	0.25	0.00	0.00
	World	0.35	1.40	0.15	0.15
Buildings	OECD	1.80	0.15	0.15	
	World	5.00	0.50	0.60	
Industry	OECD		0.35	0.35	0.20
	World		1.10	2.40	0.55
Agriculture	OECD		0.30	0.20	0.30
	World		1.60	1.10	1.70
Forestry	OECD	0.01	0.25	0.30	0.25
	World	0.15	1.10	0.90	0.65
Waste	OECD	0.10	0.06	0.00	0.00
	World	0.40	0.18	0.10	0.04
All sectors	OECD	2.20	2.10	1.30	1.10
	World	6.10	7.40	6.00	4.50

Source: Based on data from Barker *et al.* (2007).

economic 'costs' of carbon management underpin the whole argument throughout the book. This book will:

- examine global carbon cycle and climate change issues affecting the built environment and examine the sources and sinks and human intervention in these;
- critically evaluate climate change impacts on the urban and built environment;
- delineate key strategies to reduce carbon in new buildings, existing buildings and cities;
- provide an understanding of the conceptual and methodological bases for conducting high quality investigations in the context of carbon management.

The book is divided into three sections: overview; strategies for a low carbon built environment; and regulations, tools and techniques for carbon management in the built environment. These sections cover the following:

- the global carbon cycle and climate change: the global carbon cycle, the biological/physical carbon cycle, and carbon on land and in oceans;
- the science of climate change, drivers, uncertainties, policy issues, and the international treaty and national legislative framework;
- energy use and the carbon cycle: trends in global energy use, energy use and economic development, energy efficiency and economic instruments, the role of fossil fuels and renewable sources of energy, sustainable energy production, barriers and incentives, and treaty obligations;
- carbon management options: reducing sources and increasing sinks of carbon from the atmosphere, reducing sources of carbon in the built environment and cities, adaptation and mitigation, enhancing energy efficiency, embodied energy, materials management, urban transportation planning, and urban vegetation management;

- policy options for carbon reduction: international policies, the EU context, UK energy policy, and non-carbon implications of energy policies (equity dimensions, security of supply, social and economic dimensions);
- state of the art in carbon management in new build, existing stock and cities;
- tools for carbon demand reduction: policy options (e.g. Kyoto, energy trading structures, UK/Scottish climate change bills), fiscal policy (e.g. climate change levy), and tools at the micro-level (carbon footprinting, energy auditing and demand management);
- carbon economics: carbon trading and carbon offsetting, and whole life valuation in the built environment.

There are several options available to the built environment sector of the economy to contribute to this goal, and the purpose of this book is to elucidate these options and ways to measure their effectiveness in specific world economic contexts. However, a note of caution is in order, especially with respect to carbon emission in the building sector. What is needed to stabilise global warming and therefore create a sustaining environment for all life on earth is the reduction in carbon (i.e. lowering the carbon intensity of the economy) rather than the energy intensity (Herring, 2009). In the context of the built environment, these two goals (energy efficiency and 'carbon efficiency') are often used interchangeably and in a confusing manner. It is worthwhile to remember that, even as energy efficiency across all of the economic sectors has increased in recent years in OECD countries, the total energy use has not fallen (cf. Herring, 2009; Pielke, 2010). Mere concentration on energy efficiency in buildings therefore will not deliver the required carbon savings (Natarajan and Levermore, 2007). What is needed is a greater emphasis on the carbon intensity of the economy as a whole and, in our case, the building sector in particular. A focus on energy efficiency (demand reduction) will lead to the so-called 'rebound effect' (Herring and Sorrell, 2008) (e.g. money saved by installing energy efficiency measures is used for carbon intensive activities elsewhere) and therefore will not help reduce the overall carbon emitted by human activities. One of the crucial consequences to the built environment of this 'rebound effect' is to consider carefully the urban context in which buildings are placed: thus our focus in Section B of this book is not only on the building scale but also on urban carbon management.

From the built environment point of view, energy export (i.e. local production of energy in buildings over and above the consumption need) will be needed to balance the carbon emissions (net-zero buildings) (Natarajan and Levermore, 2007). In a warming world with additional demand for cooling, even this may not be sufficient. Herring (2009) stated that, while energy efficiency might make it easier to achieve a 60 per cent cut in emissions by 2050 in the UK, it is not essential. A more technically and economically feasible and logistically easier option is to decarbonise energy production, rather than attempting to enhance energy efficiency of millions of houses (Herring, 2009). The past record of several energy producers gives us confidence that this can be done. The only way for the building stock to be part of the carbon solution is to move to a decentralised energy system, where there is substantial micro-generation. 'Energy efficiency must be integrated with local generation; they must be considered as partners rather than rivals' (Herring, 2009: 194).

Even this (on-site micro-generation) is not without problems. A single renewable energy installation such as the Whitelee Windfarm near Glasgow, UK (the largest on-shore wind farm in Europe) produces 322MW of electricity from 140 turbines, enough to power 180,000 homes. Two off-shore wind developments proposed for the Thames estuary would deliver the same amount of energy at about the same cost as 866,000 micro wind turbines (Lomas, 2009). A comparison of lifecycle energy requirements and global warming impact in Canada (Kabir *et al.*, 2012) found that a single 100kW wind turbine had a lifecycle energy requirement of 133.3 kJ/kWh, which is about 69 per cent less than 20 5kW turbines or 41 per cent less than five

20kW turbines. Global warming impact from a 100kW turbine was 17.8 gCO_2eq/kWh, which is around 58 per cent less than 20 5kW turbines and 29 per cent less than five 20kW turbines. Would it not be more expedient to concentrate the limited public and private resources on installing a few very large renewable energy installations rather than maintain and manage millions of micro-generation facilities? In order to make on-site micro-generation feasible, greater state intervention, a tighter regulatory framework and substantial funding to manufacturers of micro-generation devices are needed (Williams, 2010). The book addresses these issues in Chapter 4.

All things considered, the solution ought to be 'want less, make less', i.e. energy efficiency *and* renewable generation in a combination of macro- and micro-scales. Thus the built environment (with its huge potential contribution to energy efficiency improvement) has a role to play.

Ultimately, however, it is important to underscore the end goal we have in mind. A low carbon built environment is a means to an end, not the end itself. In our rapidly changing, increasingly polluted and high carbon world, it is well to keep in mind the 'end' we have in mind. The New Economics Foundation argued that the 'end' goal of infrastructure development (of which the built environment is a crucial part) is to 'better support a good life for . . . inhabitants while also respecting the environmental resource limits upon which all our lives depend' (Aked *et al.*, 2010: 3–4). This 'end' goal of a low carbon built environment must therefore depend on:

- *place happiness* – three core aspects of well-being to which the built environment can contribute:

  o   personal well-being – people's experience of life in relation to their physical and psychological well-being;

  o   social well-being – people's experience of life in relation to those around them: their community;

  o   economic and material well-being – people's experience of life in relation to the conditions and circumstances of their lives, including their physical surroundings;

- *place sustainability* – two key ways in which buildings have an environmental impact:

  o   resources used during construction or renovation;

  o   resources used across the lifetime of buildings' use (Aked *et al.*, 2010).

1.2 History of policies and protocols for carbon management

The policies and protocols for carbon management are a rich mosaic varying from national level legislation for energy and carbon management, best practice guides, and regional carbon trading mechanisms to global treaties and protocols. The global carbon management protocols are largely governed the Kyoto Protocol – an international treaty ratified by over 190 countries to reduce GHG emissions that affect the global climate (Cheng *et al.*, 2008). The Kyoto Protocol is the legal implementation mechanism to the United Nations Framework Convention on Climate Change (UNFCCC) adoption in May 1992. Opened for signature during the Earth Summit in Rio de Janeiro, Brazil, in June 1992, the UNFCCC entered into force in March 1994 when 154 countries ratified it. The Kyoto Protocol itself came into force in February 2005 (Cheng *et al.*, 2008).

The objective of the Convention is to stabilise atmospheric concentrations of greenhouse gases at 'safe' levels. Towards this end all parties to the convention have agreed to:

1 address climate change;
2 adapt to its effects; and
3 report their actions to implement the Convention (Fenhann and Hinostroza, 2011).

The Convention divides countries into two groups: Annex I parties, which consist of developed countries and economies in transition, and non-Annex I parties, which include primarily developing countries (Fenhann and Hinostroza, 2011). The governing body with implementation as well as scientific and technical interpretative responsibilities to the convention is the Conference of Parties (COP), and Table 1.2 lists the key milestones achieved by the COP to date.

The principal mechanisms to achieve the targeted emission reduction (both the legally binding targets for the Annex I countries and the non-binding agreements for the non-Annex I countries) are all 'market based' and mostly revolve around three types: the Clean Development Mechanism (CDM), Joint Implementation (JI) and International Emissions Trading (IET).

Table 1.2 Key milestones in the UNFCCC process

Conference of Party (COP)	*Location, date*	*Relevant procedural milestone achieved*
COP3	Kyoto, Japan, 1997	Agreed to a legally binding set of obligations (Annex I countries to lower their emission by approximately 5.2% below that of 1990 levels). This is expected to be achieved in 2008–12 (Cheng *et al.*, 2008). Non-Annex I countries agreed to non-binding obligations ('common but differentiated responsibilities'). Also known as the Kyoto Protocol.
COP5	Bonn, Germany, 1999	Guidelines for the preparation of national communications by parties included in Annex I to the Convention (annual inventories and national communications) – subsequently amended at several COP meetings.
COP7	Marrakech, Morocco, 2001	Finalized most of the Kyoto Protocol's operational details and set the stage for its ratification (also known as Marrakech Accords); sets forth the operational rules for the CDM, JI and IET (Cheng *et al.*, 2008).
COP8	New Delhi, India, 2002	Express linkages between climate change and sustainable development (Delhi Ministerial Declaration on Climate Change and Sustainable Development). Highlighted the equal importance of adaptation measures and those that are mitigatory in nature.
COP10	Buenos Aires, Argentina, 2004	Guidance on CDM, including the designation of verification authorities.
COP11	Montreal, Canada, 2005	Establishment of an Adaptation Fund. Launch of JI. Official launch of the Kyoto Protocol.
COP13	Bali, Indonesia, 2007	The Bali Action Plan, consisting of: recognition of the deeper cuts in emission needed to arrest climate change; and preparation of a measurable, reportable and verifiable nationally appropriate mitigation plan, including for developing countries (in the context of sustainable development).
COP15	Copenhagen, Denmark, 2009	Establishment of a Copenhagen Green Climate Fund as an operating entity of the financial mechanism of the Convention to support projects, programmes, policies and other activities in developing countries related to mitigation, including Reducing Emissions from Deforestation and Forest Degradation (REDD-plus), adaptation, capacity building, technology development and transfer – approaching USD 100 billion a year by 2020.
COP16	Cancun, Mexico, 2010	Establishment of a Cancun Adaptation Framework to enhance action on adaptation, including through international cooperation and coherent consideration of matters relating to adaptation under the Convention.

Source: UNFCCC website (unfccc.int).

1   The CDM, which was established under Article 12 of the Kyoto Protocol, enables Annex I parties to implement projects that reduce GHG emissions in non-Annex I parties in return for certified emission reductions (CERs). CDM projects also assist host parties in achieving sustainable development and in contributing to the ultimate objective of the Convention.

2   The JI mechanism is defined in Article 6 of the Kyoto Protocol, where an Annex I party with an emission reduction and limitation commitment under the Kyoto Protocol may implement an emission reduction or emission removal project in the territory of another Annex I party with an emission reduction and limitation commitment under the Protocol. The party implementing the project may count the resulting emission reduction units (ERUs) towards meeting its own Kyoto target. This country-to-country initiative has little direct bearing on the management of carbon in the built environment.

3   IET, which is set out in Article 17, provides for Annex I parties to acquire emission units from other Annex I parties and to use those units towards meeting a part of their targets. These units may be in the form of the initial allocation, or assigned amount units (AAUs), removal units (RMUs), units issued for the amount generated from domestic sink activities, CERs under the CDM, or ERUs generated through JI. Apart from the units generated by the CDM based CERs, little direct link to the built environment is seen.

Given the possibility for technology transfer, the worldwide search for lowest cost opportunities for reducing emissions, and the possibility for small scale and private sector organisations to play a part in these, the CDM offers potential for the built environment sector to reduce emissions, especially in the developing world.

1.3 Equity implications of carbon management

The first key concept in sustainable development as defined by the Brundtland Commission (WCED, 1987, and quoted at the beginning of the chapter) is the need for equity. Given the inescapable links between carbon management and sustainable development on the one hand and the need to prepare for a warmer world even as we attempt to limit our actions that emit carbon on the other, it is crucial that carbon management efforts in the built environment remain alive to the key need for equitable approaches to the problem of carbon emission. There are at least four equity implications specific to the built environment that need to be addressed:

1   current and projected energy intensity in buildings in different parts of the world;
2   inequities in building conditions within countries (leading to effects including fuel poverty);
3   concern as to who sets the standards for the building industry;
4   differential climate-related energy needs in different countries.

Energy intensity (and in the present context carbon intensity) of buildings varies hugely between countries and even within regions of a given country. There are historical reasons for such variations, and these need to be respected. Many developing countries currently use too little energy to make their buildings comfortable and/or healthy. A focus on energy efficiency alone cannot be just or fair. A one-size-fits-all approach cannot work in such a context.

There are within-country variations in building quality, fuel use, levels of comfort, and health which lead to inequitable outcomes. People in poverty are the most vulnerable to the negative effects of climate change, as they tend to have a lower level of physical and mental health, live in worse housing with less access to insurance, and have fewer resources to cope with rising costs

(Johnson, Simms and Cochrane, 2008). While a low carbon built environment could in theory be good for energy efficiency (and therefore contribute to reducing fuel poverty), care must be taken to share the burden of decarbonising the building stock equitably. Failure to do so may exasperate the fuel poverty already in place. The case of fuel poverty in developing countries, especially in urban slums, is even more pressing. The equity implications of decarbonising the built environment will be largely influenced by climate change, energy efficiency improvements to the building stock, the price of energy, and household income (cf. Dresner and Ekins, 2005).

The development and codification of global building performance standards continue to be lopsided, with developing countries and regions having little say in defining context-specific codes and specifications. An equitable approach to carbon management in the built environment needs to be cognisant of the local requirements and contexts under which buildings need to operate and deliver their intended performance goals.

The variations in the energy needs of buildings in different climatic contexts remain very large. The so-called 'problem climates' of the world (warm and humid) need large amounts of energy to cool them, and there are no effective and widely available carbon-neutral or even low carbon options to cool buildings efficiently. This reality too must inform carbon management in the worldwide built environment context. The book recognises these issues and focuses on carbon management in different climatic regions differently (see Chapters 5 and 6).

Apart from these specific equity questions faced by the building sector, one also needs to be conscious of the wider equity implications of carbon management. Significant barriers to effective carbon management include, among others, historical responsibilities for carbon emission, benefits and costs accruing to different segments of the society, and conflict between development and carbon management.

References

All websites were last accessed on 30 November 2011.

Aked, J., Michaelson, J., Steuer, N., Boyle, D., Cox, E. and Sander-Jackson, P. 2010. *Good Foundations: Towards a Low Carbon, High Well-Being Built Environment.* London: New Economics Foundation.

Barker, T., Bashmakov, I., Alharthi, A., Amann, M., Cifuentes, L., Drexhage, J., Duan, M., Edenhofer, O., Flannery, B., Grubb, M., Hoogwijk, M., Ibitoye, F.I., Jepma, C.J., Pizer, W.A. and Yamaji, K. 2007. Mitigation from a cross-sectoral perspective. In *Climate Change 2007: Mitigation: Contribution of Working Group III to the Fourth Assessment Report of the Intergovernmental Panel on Climate Change*, ed. B. Metz, O.R. Davidson, P.R. Bosch, R. Dave and L.A. Meyer. Cambridge: Cambridge University Press.

Cheng, C., Pouffary, S., Svenningsen, N. and Callaway, M. 2008. *The Kyoto Protocol: The Clean Development Mechanism and the Building and Construction Sector: A Report for the UNEP Sustainable Buildings and Construction Initiative.* Paris: United Nations Environment Programme.

Dresner, S. and Ekins, P. 2005. *Climate Change and Fuel Poverty*, PSI Discussion Paper 24. London: Policy Studies Institute.

Fenhann, J. and Hinostroza, M. 2011. *CDM Information and Guidebook*, 3rd edn. Roskilde: UNEP Risø Centre. Available at: http://cd4cdm.org/Publications/cdm_guideline_3rd_edition.pdf.

Herring, H. 2009. National building stocks: addressing energy consumption or decarbonization? *Building Research and Information*, **37**, pp. 192–195.

Herring, H. and Sorrell, S. 2008. *Energy Efficiency and Sustainable Consumption: The Rebound Effect.* Basingstoke: Palgrave Macmillan.

Johnson, V., Simms, A. and Cochrane, C. 2008. *Tackling Climate Change, Reducing Poverty: The First Report of the Roundtable on Climate Change and Poverty in the UK.* London: New Economics Foundation.

Kabir, M.R., Rooke, B., Dassanayake, G.D.M. and Fleck, B.A. 2012. Comparative life cycle energy, emission, and economic analysis of 100 kW nameplate wind power generation. *Renewable Energy*, **37**, pp. 133–141.

Kaya, Y. 1990. Impact of carbon dioxide emission control on GNP growth: interpretation of proposed scenarios. Paper presented to the IPCC Energy and Industry Subgroup, Response Strategies Working Group, Paris (mimeo).

Levine, M., Ürge-Vorsatz, D., Blok, K., Geng, L., Harvey, D., Lang, S., Levermore, G., Mongameli Mehlwana, A., Mirasgedis, S., Novikova, A., Rilling, J. and Yoshino, H. 2007. Residential and commercial buildings. In *Climate Change 2007: Mitigation: Contribution of Working Group III to the Fourth Assessment Report of the Intergovernmental Panel on Climate Change*, ed. B. Metz, O.R. Davidson, P.R. Bosch, R. Dave and L.A. Meyer. Cambridge: Cambridge University Press.

Lomas, K.J. 2009. Decarbonizing national housing stocks: strategies, barriers and measurement. *Building Research and Information*, **37**, pp. 187–191.

Natarajan, S. and Levermore, G. 2007. Domestic futures: which way to a low-carbon housing stock? *Energy Policy*, **35**, pp. 5728–5736.

Pielke, R.A. 2010. The carbon economy and climate mitigation: an editorial essay. *WIREs Climate Change*, **1**, pp. 770–772.

Prins, G., Cook, M., Green, C., Hulme, M., Korhola, A., Korhola, E.-R., Pielke, R., Sawa, A., Stehr, N. and Storch, H. von. 2009. *How to Get Climate Policy Back on Course*. Oxford and London: Institute for Science, Innovation and Society, Oxford University and London School of Economics, Mackinder Programme. Available at: http://eureka.bodleian.ox.ac.uk/92/.

Svirejeva-Hopkins, A., Schellnhuber, H.J. and Pomaz, V.L. 2004. Urbanised territories as a specific component of the global carbon cycle. *Ecological Modelling*, **173**, pp. 295–312.

UNEP (United Nations Environment Programme). 2007. *Buildings and Climate Change: Status, Challenges and Opportunities*. Nairobi: UNEP.

WCED (World Commission on Economic Development). 1987. *Our Common Future*. New York: United Nations. Available at: http://www.un-documents.net/ocf-02.htm.

Williams, J. 2010. The deployment of decentralised energy systems as part of the housing growth programme in the UK. *Energy Policy*, **38**, pp. 7604–7613.

2 Overview of climate change

Climate change is the single greatest problem facing humanity today. Its impacts will be felt by many generations to come, and the future of our entire planet will be determined by how effectively we tackle it. Time is running short, and decisions made over reducing greenhouse gases today will directly affect how much we can limit the impacts of climate change and how much we will be forced to absorb them. However, getting those decisions right requires evidence that is both understandable and based on rigorous science, and carbon accounting has a major role to play in developing this evidence. This chapter gives a brief introduction to the science of climate change, climate change impacts, the greenhouse gases, and some of the key debates in the field.

2.1 Climate change science and the greenhouse gases (GHGs)

Climate change science is the study of the significant and long term changes to the earth's climate generated by both natural cycles and the impacts of human activity (Houghton, 2001). 'Long term' can refer to anything from over a decade to millions of years, although 30 years is a commonly used averaging period. Also, different authors often use different terms to be more or less specific when discussing the subject. For example, some authors, including many prominent sceptics of climate science, use the term 'anthropogenic climate change' to specifically denote human-induced climate change; and in some parts of the world, such as the USA, 'global warming' (which specifically relates to changes in the greenhouse effect) is still sometimes used to refer to wider changes in the climate system. Therefore to clarify these subtle but important distinctions this text uses the terminology advocated by the United Nations Framework Convention on Climate Change (UNFCCC) and defines climate change as 'a change of climate which is attributed directly or indirectly to human activity that alters the composition of the global atmosphere and which is in addition to natural climate variability observed over comparable time periods', and climate change that is attributable specifically to natural causes is referred to as 'climate variability' (UNFCCC, 1992, Article 1).

The most well known of the influences on climate change is the emission of greenhouse gases (GHGs) (also termed 'forcing factors') from the burning of fossil fuels. However, other human activities that can produce a net increase of GHGs in the atmosphere include deforestation, changes in land use, and livestock farming. The main natural forcing factors influencing climate variability are volcanic eruptions and variations in solar irradiation (the amount of energy reaching earth from the sun) (Royal Society, 2010). Sceptics of climate change science frequently assert that these natural factors are still having a significant influence on climate change, but there is no authoritative evidence for this. Extreme volcanic activity leading to significant climate change certainly has happened in the distant past, but individual volcanic eruptions largely have an impact on climate only locally and for the period of a few years (Royal Society, 2010). Changes in solar irradiation are thought to have been behind some noticeable perturbations in the climate in modern human history, for example the 'Little Ice Age' of 1650 to 1850. However, it is not possible to explain how this has led to the increases in average global temperatures observed since then (NASA, n.d.).

The term GHG refers to the group of gases known to be significantly contributing to climate change, which are now commonly known as 'the Kyoto basket' after their inclusion in the Kyoto Protocol (see Chapter 1). These are given in Table 2.1, which also gives their radiative forcing and global warming potential (GWP).

Radiative forcing is a complex factor that is most simply described as 'the rate of energy change per unit area of the globe as measured at the top of the atmosphere' (IPCC, 2001). It is essentially a measure of how a gas changes in the balance of energy entering and leaving the earth's atmosphere, against a baseline of 1750 (IPCC, 2007a).

GWP is a relative measure of how much heat a gas traps in the atmosphere, and is the factor used in carbon accounting to calculate the impact of one unit of each gas compared to one unit of carbon dioxide, averaged over a 100-year period. This is often expressed as CO_2 equivalent (or CO_2e).

Water vapour is also a GHG and the most potent, but is not included in the Kyoto basket owing to the limited and localised contributions of emissions from human activities, and also its short lifespan in the atmosphere and the complexities of its roles in natural cycles. Similarly, ozone (O_3) also acts as a greenhouse gas when present in the troposphere at latitudes close to the equator, but it can also act to regulate the atmospheric lifespans of other greenhouse gases.

Attempts to understand and predict the weather are as old as humanity itself, but many people think of climate change science as something new. However, it was in 1824 that the French scientist Joseph Fourier first described the greenhouse effect (Fourier, 1824). More significantly, in 1847 the American scientist George Perkins Marsh, on whom history may well confer the honour of being the first climate scientist, made this prescient statement as part of a landmark address to the Agricultural Society of Rutland, Vermont, USA: 'But though man cannot at his pleasure command

Table 2.1 The Kyoto basket of GHGs

Greenhouse gas	Chemical symbol	Radiative forcing (Wm-2)	Global warming potential
Carbon dioxide	CO_2	1.66	1
Methane	CH_4	0.48	23 (revised from 21)
Nitrous oxide	N_2O	0.16	310
Hydrofluorocarbons	HFCs	0.0004–0.033	140–11,700
Perfluorocarbons	PFCs	0.0008–0.0034	6,500–9,200
Sulphur hexafluoride	SF_6	0.029	23,900

Source: IPCC, 2007b.

the rain and the sunshine, the wind and frost and snow, yet it is certain that climate itself has in many instances been gradually changed and ameliorated or deteriorated by human action' (Marsh, 1847).

2.2 Global greenhouse gas emissions

Global emissions of GHGs vary heavily by country and sector, and how to proportion emissions reduction targets between different countries and sectors is the subject of intense national and international debates. As described in Chapter 1, emission targets and current progress towards their reductions are discussed at the COPs, the highest negotiating meetings at the global level. However, despite the Kyoto Protocol expiring at the end of 2012, no acceptable replacement protocol for setting national GHG reduction targets has been formulated.

One of the most significant barriers to formulating international agreements on climate change centres around the determining of a 'safe' and 'acceptable' figure for the average global temperature rise that we will be able to adapt to, and then developing mitigation strategies to limit emissions to a level that stands a better-than-average chance of achieving that aim. The problem with this is that, according to the work of climate scientist James Hansen of NASA, and supported by many other world-leading experts, stopping average global temperature rise at the proposed 'safe' limit of 2°C by 2100 will require limiting the amount of CO_2 in the atmosphere to 350 parts per million (ppm), which is almost 30 ppm below the current level of around 387 ppm (Hansen *et al.*, 2008).

National levels of GHG emissions are commonly expressed as totals and per capita, and these figures are a major focus of political negotiations. Table 2.2 shows how China, whilst having the highest total emissions for 2005, has much lower per capita emissions than the USA, and the difference is pronounced for many developed countries such as Germany and the UK. However, the country with the highest per capita emissions for 2005 was Qatar, which achieved a per capita output of 55.5 $mtCO_2$ (note this is CO_2 only). Negotiations over apportioning emissions reduction targets frequently focus on how much each country should agree to decrease its emissions, or be allowed to increase them, according to this and other indicators of its level of development. This process is sometimes known as 'contraction and convergence'. However, this term has a specific origin and meaning (see 2.2.1 below).

Total and per capita emissions are further confounded by the differences in historical contributions to climate change. Figure 2.1 shows the share of global CO_2 emissions for the top five contributors for 1850 to 2002, over which period the developed world has been responsible for around 76 per cent of the overall total. However, because our knowledge and awareness of climate change are relatively recent (and robust measurements of greenhouse gases are even more so), those most responsible historically invariably argue that these emissions should be discounted, with reduction commitments usually based on baselines of 1990 or 1995.

Table 2.2 Total and per capita CO_2e emissions in selected countries

Country	Total emissions (GtCO2e)	Per capita emissions (tCO2e)
China	7.22	5.5
USA	6.95	23.5
Brazil	2.86	5.4
Russian Federation	2.02	13.7
India	1.88	1.7
Germany	1.01	11.9
United Kingdom	0.69	10.6

Source: Based on data available from WRI, 2011.

Note: Figures for per capita emissions exclude those from land use change.

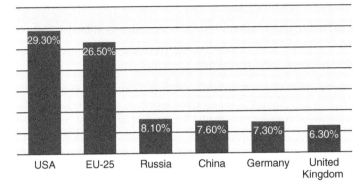

Figure 2.1 Cumulative CO$_2$ emissions as a percentage of world emissions, 1850–2002
Source: Baumert *et al.*, 2005.

The main source of global GHGs, both current and historical, is the burning of fossil fuels. However, reducing the contribution from cement production is a key research and development priority for the construction industry (see Table 2.3).

Another issue for decision makers at many levels is the varying contributions of greenhouse gases from different sectors and activities. Globally, electricity and heat production is responsible for the largest share of emissions, followed by industry, transportation, agriculture and land use change. However, the contributions of these sectors vary internationally, as does their potential contribution to reducing emissions (see Figure 2.2).

Many of the difficulties in agreeing emissions reduction targets, as well as how much to prioritise mitigating climate change as opposed to adapting to it, rest on the issue of equity as discussed in Chapter 1. Historically, the developed nations of the world have contributed the most to climate change, but now China, India and other rapidly developing countries have become significant contributors, and all argue that they should be able to increase their emissions as they develop, whilst the developed nations begin to pay off their emission debts. A further problem for ensuring equity is that the earliest major impacts of climate change are largely affecting nations with lower historical emissions and higher levels of poverty, and, although there is international agreement that the developing countries should be allowed to increase their emissions in order to develop, there is very little agreement over how much by and for how long.

2.2.1 Contraction and Convergence

One of the most widely advocated and scientifically sound models for resolving this problem of reducing global emissions whilst ensuring greater equity is Contraction and Convergence™

Table 2.3 Main sources of CO$_2$ from fossil fuels, 2000–04

Source	Contribution (%)
Liquid fuels (e.g. petrol, oil)	36
Solid fuels (e.g. coal)	35
Gaseous fuels (e.g. natural gas)	20
Transport sources not included in national inventories	4
Cement production	3
Flaring gas from wells and industry	< 1
Non-fuel hydrocarbons	< 1

Source: Raupach *et al.*, 2007.

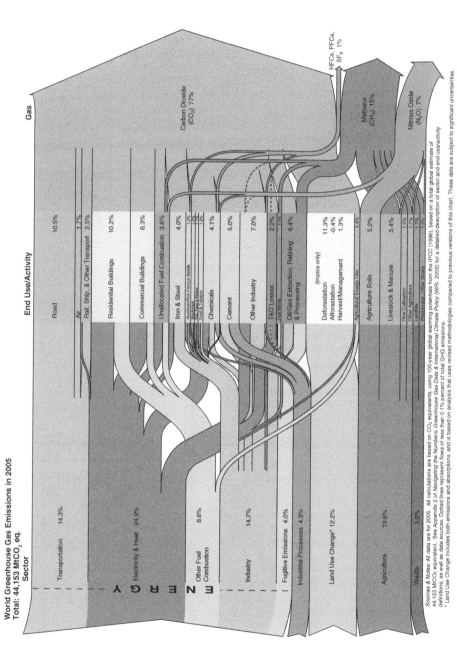

Figure 2.2 World GHG emissions by sector, 2005

Source: WRI, 2011.

(Meyer, 2001). C&C, as it is known, was originally developed by Aubrey Meyer of the Global Commons Institute (GCI). However, the term has been adopted more widely, and where it is used it is important to know whether or not the specific model is being referred to. C&C begins with the principle that the developing world should be allowed to develop whilst the developed world begins to reduce its emissions, and then models these trajectories over time to meet emissions goals of 350 ppm, 450 ppm and 550 ppm. The best way to understand C&C is to inspect the highly zoom-able diagram produced by the GCI. Figure 2.3 gives a snapshot of the diagram, which is free to download from the GCI website.

2.3 Greenhouse gas sources

As previously discussed, GHGs are emitted from both natural sources and human activity. The primary task of carbon management is to measure anthropogenic emissions and justifiably attribute them to their sources.

Globally, the main anthropogenic sources of greenhouse gases are (see also Figure 2.2 for details):

- burning fossil fuels;
- deforestation;
- land use and wetland changes;
- livestock enteric fermentation and manure management;
- paddy rice farming;
- pipeline losses;
- emissions from landfill sites;
- use of CFCs in refrigeration systems;
- CFCs and halons in fire suppression systems and manufacturing processes;
- agricultural activities, such as the use of fertilisers (IPCC, 2007b).

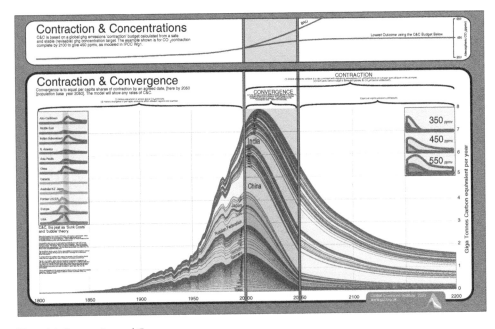

Figure 2.3 Contraction and Convergence
Source: Meyer, 2011.

Of these, the burning of fossil fuels is by far the most significant. However, it is also (relatively) the easiest to measure accurately. Reducing, and ultimately eliminating, the consumption of fossil fuels is the top priority for reducing emissions. Nevertheless, meeting global emissions reduction targets will require all these other sources to be addressed.

Within fossil fuels, our dependency on oil is the greatest concern. Experts argue over whether or not we have passed 'peak oil' (see 2.3.1 below), and if not when we will pass it, but there is also the question of how much we should be burning in light of its other uses. However, perhaps the most significant concern is coal, which is undergoing a revival following the development of carbon capture and storage (CCS) and coal gasification technologies. For many countries meeting demand that cannot be met from renewable energy sources, CCS-equipped fossil fuel (and large-scale biomass) plants may be an attractive option compared to nuclear. However, it remains technologically and economically unproven at a sufficiently large scale.

2.3.1 Peak oil

The theory of peak oil was first coined by M. King Hubbert, a geophysicist working for Shell, in 1956, and is widely accepted by geologists (Hubbert, 1956). Hubbert theorised that as the global resource of oil is finite (except at geological timescales) then the rate at which oil extraction is increasing should peak at some predictable point in time and then begin a terminal decline. Two caveats to the theory are that it applies to the reserves of 'easily extractable' oil (i.e. that which can be extracted using conventional drilling techniques) and that an accurate prediction of the peak requires accurate data on the existing volume of oil available for extraction. Reserves are distinguished from resources by being extractable using techniques that are both currently available and commercially viable. However, the factors that determine whether any given deposit is a reserve or a resource are broader and open to degrees of uncertainty. For example, whilst oil extraction from tar sands in Canada is now a technologically and commercially viable process, the volume of water used (2–4.5 barrels of water to 1 barrel of oil in 2008) poses a resource limitation that, arguably, means tar sands should not be classified as reserves. The theory quickly gained traction when Hubbert's prediction that US oil extraction would peak in 1970 came true, albeit at a higher output than forecast, and, although it remains a contested theory, peak oil has become established in mainstream thinking on energy and climate change (Hubbert, 1956; Verbruggen and Al Marchohi, 2010).

Following on from peak oil, researchers have applied the theory to predict peak production of other finite resources such as coal, gas, uranium, and the rare earth metals used in electronic goods.

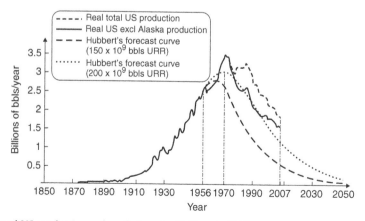

Figure 2.4 Actual US production and predictions by Hubbert in 1956

Source: Verbruggen and Al Marchohi, 2010.

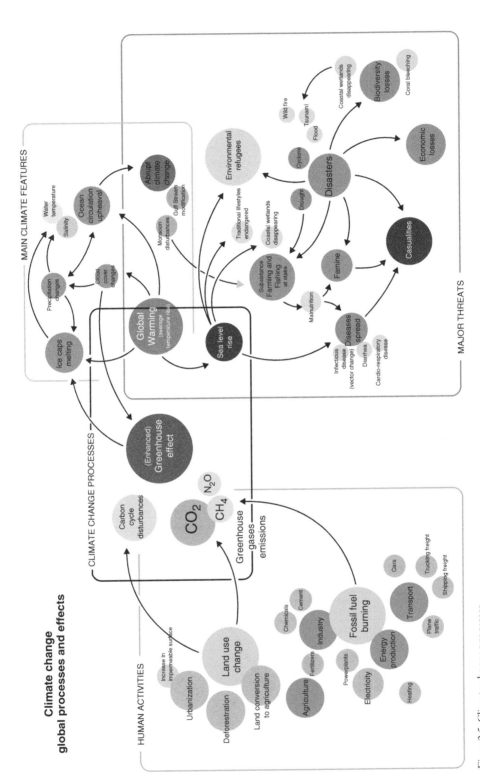

Figure 2.5 Climate change processes

Source: UNEP/GRID – Arendal, 2009.

2.4 Greenhouse gas sinks

Greenhouse gases can be removed from the atmosphere by various processes, including: physical changes in the atmosphere, such as precipitation; photosynthesis; absorption by the oceans; and chemical and physical reactions within the atmosphere, such as the natural oxidation of methane and the dissociation of halocarbons by UV light. As yet there are few manufactured technologies capable of sequestering greenhouse gases from the atmosphere, with the most prominent being the production of biochar, carbon capture and storage, carbon dioxide air capture, and ocean seeding. However, all are controversial, and with the partial exception of ocean seeding these involve the eventual burial of sequestered carbon dioxide.

2.5 Adaptation and mitigation

In a world of limited resources an important question for decision makers is how much to invest in mitigating future climate change and how much to invest in adapting to it. This is difficult, because the greater the amount of emissions we generate the greater the risk of climate change that we commit to adapting to in the future, but owing to the complexity of the problem we can only estimate the severity and extent of those impacts.

The most authoritative source on the predicted future impacts of climate change is the Fourth Assessment Report completed by the Intergovernmental Panel on Climate Change (IPCC) in 2007 (IPCC, 2007a), which estimates that globally averaged surface temperatures will be 2.5–4.7°C higher by 2100 compared to pre-industrial levels, with the full range of uncertainty in projected temperature increases by 2100 being 1.8–7.1°C depending on various scenarios and uncertainties in climate sensitivity (Solomon *et al.*, 2007). Reducing emissions to meet different

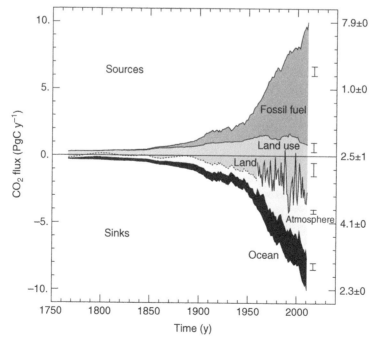

Figure 2.6 Human perturbation to sources and sinks of global carbon, 2000–01
Source: Canadell, 2011 (based on data from Le Quéré *et al.*, 2009; Canadell *et al.*, 2007),.

'acceptable' temperature targets is predicted to have significantly different outcomes for the planet, which further underlines the importance of not only agreeing emissions reduction targets but also having detailed and realistic plans for achieving them.

2.6 Vulnerability

How at risk a country, ecosystem or resource (and so on) is to the impacts of climate change is known as its vulnerability. The 'double whammy' of vulnerability is that those parts of the world at greatest immediate risk from severe impacts of climate change tend to be in the tropics and at the poles, which also contain some of the world's poorest nations and most at-risk ecosystems. However, even in the developed world many low-lying countries, such as the Netherlands, and major cities, such as London and New York, are having to invest heavily in protecting themselves against sea level rise.

In 2011, the ten most vulnerable countries were as follows:

1   Bangladesh;
2   India;
3   Madagascar;
4   Nepal;
5   Mozambique;
6   the Philippines;
7   Haiti;
8   Afghanistan;
9   Zimbabwe;
10 Myanmar.

A problem for linking vulnerability to climate change is understanding how much influence climate change is having on the frequency and strength of extreme weather events such as hurricanes and flooding, the risk of which is itself a component of vulnerability. It is also necessary to assess the likelihoods of two or more risks occurring at the same time, and what the cumulative impacts of these would be.

Although it is likely that some risks will be allowed to occur, not protecting the most vulnerable countries from the impacts of climate change now will only cause greater problems in the future as climate change drives population migrations – and in the case of some island countries such as the Maldives this will mean moving their entire population.

Bibliography

All websites were last accessed on 30 November 2011.

Baumert, K.A., Herzog, T. and Pershing, J. 2005. *Navigating the Numbers: Greenhouse Gas Data and International Climate Policy*. Washington, DC: World Resources Institute. Available at: http://pdf.wri.org/navigating_numbers.pdf.

Canadell, J. 2011. *Carbon Budget 2010*. Canberra: Global Carbon Project. Available at: http://www.csiro.au/en/Portals/Multimedia/CSIROpod/~/media/CSIROau/Publications/CarbonBudget2010.pdf.

Canadell, J.G., Le Quéré, C., Raupach, M.R., Field, C.B., Buitenhuis, E.T., Ciais, P., Conway, T.J., Gillett, N.P., Houghton, R.A. and Marland, G. 2007. Contributions to accelerating atmospheric CO_2 growth from economic activity, carbon intensity, and efficiency of natural sinks. *Proceedings of the National Academy of Science*, **104**, 20 November, pp. 18866–18870. Available at: http://www.pnas.org/content/104/47/18866.abstract.

Fourier, J. 1824. Remarques générales sur les températures du globe terrestre et des espaces planétaires. *Annales de Chimie et de Physique*, **27**, pp. 136–167.

Hansen, J., Sato, S., Kharecha, P., Beerling, D., Masson-Delmotte, V., Pagani, M., Raymo, M., Royer, D.L. and Zachos, J.C. 2008. *Target Atmospheric CO₂: Where Should Humanity Aim?* New York: NASA/Goddard Institute for Space Studies. Available at: http://www.columbia.edu/~jeh1/2008/TargetCO2_20080407.pdf.

Houghton, J.T. (ed.). 2001. Appendix I: glossary. In *Climate Change 2001: The Scientific Basis: Contribution of Working Group I to the Third Assessment Report of the Intergovernmental Panel on Climate Change.* Cambridge: Cambridge University Press. Available at: http://www.grida.no/publications/other/ipcc_tar/?src=/climate/ipcc_tar/wg1/.

Hubbert, M.K. 1956. *Nuclear Energy and Fossil Fuels*, Publication no. 95. Houston, TX: Exploration and Production Research Division, Shell Development Company.

IPCC. 1995. *IPCC Second Assessment: Climate Change, 1995: A Report of the Intergovernmental Panel on Climate Change.* New York: United Nations. Available at: http://www.ipcc.ch/pdf/climate-changes-1995/ipcc-2nd-assessment/2nd-assessment-en.pdf.

IPCC. 2001. *Third Assessment Report: Climate Change 2001: Working Group 1: The Scientific Basis.* Geneva, IPCC. Available at: http://www.grida.no/publications/other/ipcc_tar/?src=/climate/ipcc_tar/wg1/222.htm.

IPCC. 2007a. *Climate Change 2007: Synthesis Report.* Geneva, IPCC. Available at: http://www.ipcc.ch/pdf/assessment-report/ar4/syr/ar4_syr.pdf.

IPCC. 2007b. Summary for policymakers. In *Climate Change 2007: Impacts, Adaptation and Vulnerability: Contribution of Working Group II to the Fourth Assessment Report of the Intergovernmental Panel on Climate Change*, ed. M.L. Parry, O.F. Canziani, J.P. Palutikof, P.J. van der Linden and C.E. Hanson, pp. 7–22. Cambridge: Cambridge University Press. Available at: http://www.ipcc.ch/pdf/assessment-report/ar4/wg2/ar4-wg2-spm.pdf.

Le Quéré, C., Raupach, M.R., Canadell, J.G., Marland, G. *et al.* 2009. Trends in the sources and sinks of carbon dioxide. *Nature Geosciences*, **2**, pp. 831–836. Available at: http://www.nature.com/ngeo/journal/v2/n12/full/ngeo689.html.

Marsh, G.P. 1847. Address delivered before the Agricultural Society of Rutland County, Sept. 30, 1847. By George P. Marsh. Published by the *Herald*, Rutland, VT, in 1848. Available at: http://memory.loc.gov/cgi-bin/query/r?ammem/consrv:@field%28DOCID+@lit%28amrvgvg02div1%29%29.

Meyer, A. 2001. *Contraction and Convergence: The Global Solution to Climate Change*, Shumacher Briefings no. 5. Totnes: Green Books.

NASA. n.d. *Global Climate Change: Causes.* Washington, DC: North American Space Administration. Available at: http://climate.nasa.gov/causes/.

Raupach, M.R., Marland, G., Ciais, P., Le Quéré, C., Canadell, J.G., Klepper, G. and Field, C.B. 2007. Global and regional drivers of accelerating CO_2 emissions. *Proceedings of the National Academy of Sciences of the USA*, **104** (24), pp. 10288–10293.

Royal Society. 2010. *Climate Change: A Summary of the Science*, September. London: Royal Society. Available at: http://royalsociety.org/climate-change-summary-of-science/.

Solomon, S., Qin, D., Manning, M., Chen, Z., Marquis, M., Avery, K.B., Tignor, M. and Miller, H.L. (eds). 2007. *Contribution of Working Group I to the Fourth Assessment Report of the Intergovernmental Panel on Climate Change, 2007.* Cambridge: Cambridge University Press. Available at: http://ipcc.ch/publications_and_data/ar4/wg1/en/contents.html.

UNEP/GRID-Arendal. 2009. *Climate Change Global Processes and Effects.* Arendal: UNEP/GRID-Arendal Maps and Graphics Library. Available at: http://maps.grida.no/go/graphic/climate-change-global-processes-and-effects1.

UNFCCC. 1992. *The United Nations Framework Convention on Climate Change.* Bonn: United Nations Framework Convention on Climate Change. Available at: http://unfccc.int/essential_background/convention/background/items/1349.php.

UNFCCC. 2007. *Climate Change: Impacts, Vulnerabilities and Adaptation in Developing Countries.* Bonn: United Nations Framework Convention on Climate Change. Available at: http://unfccc.int/resource/docs/publications/impacts.pdf.

Verbruggen, A. and Al Marchohi, M. 2010. Views on peak oil and its relation to climate change policy. *Energy Policy*, **38**, pp. 5572–5581.

World Resources Institute (WRI). 2011. *Climate Analysis Indicator Tool.* Washington, DC: WRI. Available at: http://cait.wri.org/.

3 Sectoral approaches to carbon management

As discussed in Chapter 2, reducing greenhouse gas emissions requires action across all sectors of the economy and society, and the built environment provides many challenges and opportunities for doing so. The specific challenges faced by the built environment and strategies to address carbon in new build, existing stock and cities are presented in Section B. This chapter summarises the key sectors and other areas where there is significant potential to reduce the emissions from built environments and the infrastructures that underpin them. However, it is important to remember that there is a high level of interactivity between all of these systems, and the most effective strategies to reduce emissions will be those that account for these and capitalise on any opportunities to use packages of measures to deliver greater cumulative outcomes.

3.1 Energy generation

The debate over how to reduce emissions from energy generation has, until recently, largely been focused on the fuel source. However, and aside from concerns over the wider impacts of major developments, the growth and diversity of renewable technologies and increasing concerns over energy security have seen this debate expand to include where that energy is being generated, and how (and how far) it is distributed. Yet this debate, which can be traced back to George Westinghouse's victory over Thomas Edison in the 'war of the currents', was fundamental in establishing the predominance of the system of centralised distribution that most of us still rely on today (McNichol, 2006). An often overlooked outcome of this war is that, had Edison won, decentralised micro-generation using direct current (DC) networks may still have been common today.

Edison is often (erroneously) credited with being the father of the electric light bulb, but it is his original vision of how electricity should be generated and distributed that may be his most prescient legacy. This vision was one of businesses generating their own electricity on site and selling the excess to domestic users via a local grid, for which he invented a meter to measure consumption, which worked only for DC. Indeed, the use of low voltage DC networks in some urban areas continued long after he conceded defeat to Westinghouse, with Stockholm operating one until as late as the 1970s (Blalock, 2006). However, and somewhat ironically, it was the invention of an electricity meter suitable for alternate current (AC) (along with the transformer and the electric motor) that ultimately handed the victory to Westinghouse and his business partner, Croatia's 'troubled genius' Nikolai Tesla.

The legacy of this 'war' has huge implications for our attempts to reduce emissions from the energy sector. The major advantage of high voltage AC distribution is its significantly lower transmission losses, but the infrastructure needed for this favours large centralised generation technologies such as fossil fuel plants, hydro, nuclear, and large scale renewable installations. It also makes it technically difficult (but far from impossible) to re-engineer national electricity grids to allow micro-generation systems to feed excess electricity into them, thereby enabling the balancing of the changing levels of electricity being supplied to them (Wissner, 2010). Another crucial factor favouring the status quo is the need to supply a 'baseload' level of electricity for heavy industry and essential services – a common argument used by proponents of nuclear power as a low carbon electricity source. Finally, there is the problem of the cost of restructuring electricity grids and who should bear it, the uncertainty over which has been found to be a particularly significant barrier to both restructuring and the deployment of renewables. However, examples from countries including Germany and the Netherlands are now demonstrating that none of these barriers are insurmountable (Swider *et al.*, 2008).

Those now working in the energy sector are facing a range of often conflicting demands. Most importantly, national and global emissions reduction targets invariably rely on predictions regarding the 'decarbonisation' of energy supply, i.e. the amount of future demand that will be met by low or zero carbon energy sources. If decarbonisation targets are not met, or not met in time, then there is little hope of achieving the cumulative targets. For example, achieving Scotland's national target for 2050 will require the country's net emissions to fall below those currently emitted by its energy sector alone (see Figure 3.1), much of which depends on the contribution from grid decarbonisation (Scottish Government, 2009a).

The need to ensure energy security and/or energy independence is another key demand, and one which poses opportunities and barriers for the large scale deployment of renewables. On

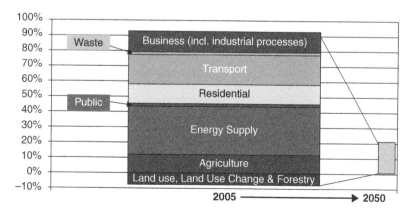

Figure 3.1 Total net GHG emissions in 2005 in Scotland by sector, compared to the target for total net emissions in 2050

one hand, many countries have the potential to generate most or all of their energy demand from renewables (Resch *et al.*, 2008), but doing so requires upgrading and expanding the energy infrastructure to allow for the variability in supply from different sources, and to reach geographically remote areas suitable for wave power and wind and solar farms. Attempts to do so also frequently run into opposition from conservationists and other lobby groups. Similarly, expanding the deployment of small and micro renewables in more urban areas requires re-engineering the existing (often deeply embedded) energy infrastructure, and is also often hampered by planning and conservation legislation and local opposition. However, increasing concerns over issues such as peak oil (see Chapter 2) and the security of gas and oil pipelines running through disputed territories may mean that grid restructuring and renewables should be at the heart of national strategies aimed at ensuring future energy security.

In addition, it is necessary to consider the carbon cost of heating our built environment, much of the demand for which is met by natural gas (nearly 40 per cent of the total in the UK; see Table 3.1). In the short term natural gas is likely to continue to make a significant contribution to global energy demand, but as a finite resource that comes with its own carbon cost and energy security concerns it cannot be relied upon in the long term. At present, shifting heating demand away from natural gas often means meeting it using electricity, but the take-up of alternatives is increasing. At a building level, solar thermal is an established technology that can be highly effective in meeting heating and hot water demands (Allen, Hammond and McManus, 2008), whilst at a larger scale the potential for supplying heat to more densely populated urban areas through combined heat and power (CHP) and district heating schemes has long been demonstrated in countries such as Denmark (Torekov, Bahnsen and Qvale, 2007). However, these are hampered by the problems of installing the necessary infrastructure, and they also have a carbon cost associated with their fuel source, which is commonly gas or biomass. Domestic CHP units are becoming increasingly available (De Paepe, D'Herdt and Mertens, 2006), but they too have their drawbacks, principally noise and vibration (Allen, Hammond and McManus, 2008).

Other renewable energy and heat options that could be deployed in urban or suburban areas include deep geothermal, energy from waste, biofuels and advanced anaerobic digestion – each with different advantages and disadvantages. The differing rates at which these options are taken up, their output and their geographic location will all be interdependent on future strategies for distribution.

Table 3.1 Electricity flow in the UK, 2010

Supply		End use	
Category	Amount (TWh)	Category	Amount (TWh)
Import	7.1	Energy industry own use	28.6
Solar, wind and wave	10.2	Conversion, T&D losses	568.7
Hydro	6.8	Exports	4.5
Nuclear	162.2		
Fossil fuels			
Petroleum	13.6	Iron and steel	3.5
Thermal renewables	51.4	Other industries	101.0
Manufactured fuels	9.5	Transport	3.9
Natural gas	371.7	Domestic	118.7
Coal	297.3	Other consumers	101.2
Total	929.8	Total	930.1

Source: Based on data from DECC, 2011.

This complex mix of demands, drivers, barriers and opportunities means that, in contrast to the 'current war', today's 'energy war' is unlikely to have a clear winner, but the outcome could still be decided by the options we invest in for the future of distribution. At one end of the scale the benefits of large renewable installations are driving the expansion of infrastructure both nationally and internationally, for example the European Supergrid, whilst at the other the growth in micro-generation is also driving change at local levels. The question remains as to how these complementary and yet competing options will influence the strategies adopted by governments and the energy sector.

Finally, it is worth remembering that around the world many homes, and some businesses, are opting to go 'off-grid' and meet their energy needs without connecting to the grid. Although in many parts of the world this reversion to Edison's dream is generally unfeasible in urban areas, it is also the starting position of many people in the developing world who have yet to achieve significant electrification. As these countries continue to develop, and as energy demand from the developed world continues to increase, the most significant challenge to a low carbon energy future will be meeting, and ultimately reducing, global energy demand.

3.2 Transport

Accounting for emissions from transport poses a particularly difficult set of problems for carbon accountants, yet global reliance on oil as the primary fuel source means that overcoming these is an essential step in the shift to a low carbon economy. It is relatively straightforward to estimate accurately the emissions from most forms of transport. However, the problems lie in how to attribute those emissions across administrational boundaries – otherwise known as 'trans-boundary issues' (Bruvoll and Fæhn, 2006) (see also Chapter 7). These feature most prominently in debates over aviation and shipping, but they can occur whenever a transport network links more than one organisation required to report its emissions, or when an action in one area affects traffic in another.

Compared to energy generation, the options for reducing emissions from transport by switching fuel source are relatively limited and contested. Biofuels, although hugely controversial, are currently the only viable renewable alternatives for fossil fuels or electricity derived (largely) from coal or natural gas (Blakey, Rye and Wilson, 2011), although a Japanese company is developing solar sails that could be an alternative to biofuels for shipping (*ETN*, 2011). On land, biofuels can also have their benefits if used sustainably, for example bioethanol in Brazil (de Brito Cruz, 2008) and biodiesel made from used frying oil in Bosnia and Herzegovina (Mercy Corps, 2009). Similarly some organisations have now converted fleets of vehicles to run on biogas, for example the bus services in Stockholm in Sweden and Lille in France (Stromberg, 2004).

The main alternative to biofuels is switching to electric vehicles, a policy being pursued aggressively by countries such as the UK (see London's attempt to promote electric vehicles in Chapter 7). Electric vehicles come in many forms, including those powered by simply charging from the grid, or fuel cells, hybrid vehicles, and those equipped with energy recovery systems. Three key barriers for electric vehicles are their performance when compared to conventional vehicles, the weight and sizes of their batteries or fuel tanks, and the general lack of infrastructure for refuelling or recharging (Shukla, Pekny and Venkatasubramanian, 2011), so as for the energy sector there are strong interdependencies between fuel and distribution. However, the distributed infrastructure needed to support urban transport networks also provides many opportunities for expanding micro-generation, as evidenced by the installation of solar photovoltaic (PV) panels on bus shelters and parking meters in many cities around the world.

Another effective way of reducing emissions from transport is by driving a modal shift towards greater use of public transport, cycling and walking, and shifting freight traffic to lower carbon alternatives (see Chapter 7 for attempts by many cities to promote energy efficient transport). However, enabling this shift requires re-thinking the way urban environments are planned to emphasise sustainable mobility over the conventional engineering-driven approach, as set out in Table 3.2. Furthermore, achieving this change is a multi-faceted problem that relies not only on improvements in urban design and infrastructure, but also effective behaviour change, which requires making the alternative an attractive proposition (Tiwari, Cervero and Schipper, 2011). For more on behaviour change see 3.8 below.

Transport and energy networks are the most visible of the inter-connected infrastructures that sustain human life. However, they are far from the only problem for those designing a low carbon economy. Human life and settlements also depend on infrastructure networks for water (and wastewater), waste collection and disposal, information and communications technology (ICT), and manufacturing and distribution, as well as green networks of open spaces that serve as recreational facilities and provide important habitats for wildlife (so-called 'green corridors') and often local food. These too need to be decarbonised.

3.3 Water and wastewater

The impacts of poorly designed and managed water infrastructure are most commonly associated with leaking pipes and the traffic congestion caused by digging roads to maintain or replace them. However, distributing fresh water and collecting and treating wastewater have significant energy costs. In the UK the water industry is responsible for 1 per cent of national GHG emissions, with the target of reducing these by 60 per cent by 2050. Measures to reduce these emissions include:

- reducing energy use (electricity and other fuels) through energy efficiency measures;
- water efficiency and leakage control measures;
- research and development into alternative low carbon technologies and 'soft' solutions to achieving water quality standards;

Table 3.2 Contrasting approaches to transport planning

Conventional approach: transport planning and engineering	*Alternative approach: sustainable mobility*
Physical dimensions	Social dimensions
Mobility	Accessibility
Traffic focus, particularly on the car	People focus, either in (or on) a vehicle or on foot
Large in scale	Local in scale
Street as a road	Street as a space
Motorised transport	All modes of transport, often in a hierarchy with pedestrians and cyclists at the top and car users at the bottom
Forecasting traffic	Visioning on cities
Modelling approaches	Scenario development and modelling
Economic evaluation	Multi-criteria analysis to take account of environmental and social concerns
Travel as a derived demand	Travel as a valued activity as well as a derived demand
Demand based	Management based
Speeding up traffic	Slowing movement down
Travel time minimisation	Reasonable travel times and travel time reliability
Segregation of people and traffic	Integration of people and traffic

Source: Bannister, 2008; Marshall, 2001.

- embedded renewable power generation (e.g. on pumping stations);
- purchasing or generating green electricity, and integrating combined heat and power (CHP);
- investing in technologies with lower whole life carbon impacts and costs;
- working with customers and the wider supply chain to encourage low carbon behaviour (Water UK, 2009).

Increasing demand, vulnerability of supply, more stringent quality requirements and the costs and complexities of installing and maintaining water infrastructure are proving useful drivers for carbon reduction (CST, 2009), with guidance available from sources such as Ainger *et al.* (2008).

Changing wastewater processing methods can also make a significant contribution to reducing emissions from the sector, with anaerobic digestion and membrane bioreactors being the most effective solutions. However, ultimately some solid waste is still produced from sewage, and although some operators are treating this and converting it into a fertiliser this may be unpalatable to some people, and the overriding aim should be to dispose of it into environments with the capacity to absorb it safely (EA, 2009).

3.4 Waste management

The over-arching goal for waste management is to maximise the volume of waste recovered and recycled (or converted to energy) per unit of GHG emissions emitted. Much of this depends on the efficiency of the infrastructure and the method(s) used for collection, transportation, sorting and processing. So there may be trade-offs to be made between the stages in limiting emissions, for example when calculating the emissions savings from using a higher efficiency processing plant located further from the waste supply versus a nearer but less efficient plant.

The overriding goal of waste management is to prevent, or at least minimise, the production of waste. Beyond this the challenge is to reuse and recycle as much of the waste collected as possible, and then to recover energy from what cannot be recycled (using waste to energy plants) before consigning the remainder for disposal at landfill sites (as shown in Figure 3.2).

It is a sad testament to humanity that one of the most visible man-made environments from outer space is the Fresh Kills landfill site in New York, and landfill sites themselves are a major source of GHG emissions. When accounting for the vast volumes of waste still disposed of in landfill sites it is necessary to consider the emissions generated from biodegradation, as well as the environmental impacts such as groundwater pollution. In the developed world there is no reason why any food and other organic waste, which produces methane during decomposition, should continue to reach landfill sites. Smaller volumes of organic waste can be composted at source or at community composting facilities (where they exist), whilst larger volumes can be sent to in-vessel composters or anaerobic digestion plants, the latter having the benefit of producing renewable energy (Ciotola, Lansing and Martin, 2011). However, in order to be commercially viable these need to be fed regularly with sufficient volumes of waste. Options for achieving this include increasing the number of organisations and individuals using the facility, and collecting waste from a wider geographical area. Again, the more efficient the collection system, the more efficient the plant is. However, as shown in Figure 3.3, anaerobic digestion plants are widely adaptable, making them an attractive and effective option for disposing of waste.

Many studies (e.g. Sidique, Lupi and Joshi, 2010; Gellynck, Jacobsen and Verhelst, 2011) have shown the importance of simplifying the routes between the source and processing plant. Put simply, the more sorting is required at source and the further waste has to travel to a collection bin, the less likely it is to be recycled. Using this rule of thumb it should, for example, be possible to increase the volume of material recycled by an average office by ensuring that recycling bins

Figure 3.2 The waste hierarchy

Source: SEPA, 2011.

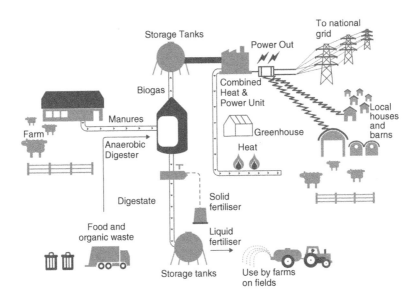

Figure 3.3 Potential for anaerobic digestion

Source: DECC, 2009 (Image processed by David Fulford of Kingdom Bioenergy Ltd, davidf@kingdombio.com).
NB: Does not show the additional uses of generating gas or transport fuel.

are positioned alongside all general waste bins, and limiting the use of general waste bins to areas generating higher volumes of mixed waste. This explains both the evidence for the recycling rates for domestic kerbside collection (Perrin and Barton, 2001) and the growth in the development of integrated waste management schemes utilising plants that mechanically sort mixed waste streams (Emery *et al.*, 2007). It is also important to design recycling systems around local factors, such as housing types and other aspects of the built environment, in order to facilitate behaviours that maximise recycling rates (Martin, Williams and Clark, 2006). A more innovative approach, pioneered in Scandinavia, is vacuum tube recycling systems (Envac, 2009). These make use of a scaled-up version of the tube systems used by US banks (and others) to transport waste to a local collection facility, thereby avoiding the emissions generated by collection vehicles going door to door. However, they are generally unsuitable for retrofitting into existing buildings.

Finally, another important factor is the types and mixes of waste involved, particularly those waste streams containing hazardous materials and/or those containing precious metals and other materials that can generate significant profits. Here it is not only waste managers that have a role to play in reducing emissions from waste management. Many products can be engineered to optimise the volume of materials that can be recovered without the need for mechanised processing and favouring the use of more homogeneous materials over more complex compounds – otherwise known as 'design for recycling'. In the EU, major drivers for innovation in this area are the Energy Using Products (EuP) Directive (EU Directive 2005/32/EC) and the Ecodesign Directive (EU Directive 2009/125/EC), and it is unsurprising that the electronics industry, which uses large volumes of valuable materials such as rare earth metals, has been quick to capitalise on the opportunities created by this legislation.

3.5 Information and communications technology

The global dependency on ICT networks presents a double-edged sword for reducing emissions. The growth of ICT networks is enabling energy efficiency in buildings and infrastructure through technologies such as smart meters and intelligent energy management systems, the latter offering the ability to control remotely the status of domestic appliances and IT systems in offices. Furthermore, many ICT networks are now being converted to being powered by direct current (DC) electricity supplied through the low voltage ethernet cables used for connecting to the internet. Ironically, this new technology is becoming available just as the few remaining older low voltage DC networks have been shut down – with Stockholm closing its network during the 1970s and Consolidated Edison closing the remains of its US network in 2007. Yet another twist in the fate of Edison's original vision is that the New Yorker Hotel, which converted to AC only in the late 1960s, was where his rival Nikolai Tesla spent his final days (Houston, 2011; Lee, 2007).

The installation of high speed broadband networks is also enabling workers in many sectors to work from home. However, this increased flexibility also serves to shift emissions from offices to homes, and in any individual case there will be a point at which emissions savings from reducing commuting are outweighed by those from homes that are now used (and heated or cooled) during work hours. In addition, the increased freedom afforded by homeworking enables other energy-consuming activities, such as shopping trips and taking children to school, which might otherwise be incorporated into the daily commute. Although much research remains to be conducted as to the impact of homeworking on energy consumption and emissions, some studies (e.g. Baker, 2007) have found that households where one or more occupants reported working from home have significantly higher energy consumption than those of non-homeworkers.

An alternative option is for employers and/or employees to relocate to limit the need for lengthy daily commuting, ideally to avoid the use of transport completely and enable commuting on foot or by bicycle. However, the feasibility of this depends on a wide range of factors, not least the availability and costs of homes and commercial premises, and therefore it is likely that homeworking via high speed IT infrastructure will continue to become more common in highly urbanised environments.

Finally, another issue for carbon management in ICT is the growth in 'cloud-based' services, where some or all software is accessed from energy-intensive server farms that can be located anywhere in the world. Therefore future assessments of carbon emissions will need to account for, and justify the attribution of, emissions generated from such networks.

3.6 Manufacturing and distribution

In many parts of the world, manufacturing industries have long since moved to the periphery of human settlements. This migration is less pronounced in rapidly industrialising countries such as China and India. Furthermore, the manufacture and distribution of food and goods are critical in sustaining life in urban areas. Therefore how to account most robustly and justifiably for emissions from manufacturing has huge implications for both the future of carbon accounting and the global response to climate change.

Increasingly, the food and goods consumed in the developed world are being labelled to disclose their energy efficiency and/or embodied carbon, either voluntarily or mandatorily. Examples of such labels are the EU's energy efficiency label for appliances, and the UK's Carbon Trust label, which is achieved by following the PAS2050 methodology.

For many consumer goods sold worldwide a significant component of the embodied carbon will be for emissions from the transport needed to sustain highly distributed global supply and distribution chains. This raises an important question for carbon accountants, because it means that, arguably, the more conventional *production-based* approaches to carbon accounting are not suitable for a globalised economy, as they incentivise low energy intensity economies (i.e. much of the developed world) to import more and manufacture less, whilst giving no incentives for these countries to tackle the global impacts of their consumption. Furthermore, the incentive to reduce emissions by exporting less and importing more invariably off-shores the generation of emissions to countries with more carbon intensive industrial and manufacturing sectors, resulting in the misunderstanding of emissions accounting that causes the 'blame China' attitude. The reasons behind the temptation to overlook this critical problem are exemplified by the work of Helm, Smale and Phillips (2007), who estimated that, were an alternative *consumption-based* approach to be applied in the UK, far from having reduced its emissions beyond its Kyoto target of 12.5 per cent by 2008–12 it would have *increased* its emissions by around 19 per cent for the period 1990–2003. Furthermore, their calculations (which are based only on aggregate figures) demonstrate the growth in off-shored emissions by finding that the discrepancy between the two approaches produces a difference of just 23 per cent for 1990, but 72 per cent for 2003.

3.7 Green spaces

Green spaces have a complex but essential role to play in reducing carbon emissions from the built environment and enhancing the sustainability of cities. From the point of view of conservationists green spaces provide valuable habitats and migration networks for wildlife facing the impacts of urban growth and the destruction of ecosystems in rural areas (Bennett, 2003). For urban planners they provide beauty and recreational spaces, which themselves may indirectly support emissions reductions through making walking and alternative forms of transport more popular with residents, and of course increasing vegetation provides new sinks for CO_2 emissions. Areas of vegetation also serve to absorb water and can help prevent localised flooding at times of high rainfall. However, heavily managed green spaces such as parks, ornamental gardens and even garden lawns can be highly resource intensive, and some studies have found this to the point that the net impact on GHG emissions is positive rather than negative (e.g. Townsend-Small and Czimczik, 2010).

Green spaces can also serve to reduce emissions from cities by providing space for residents to grow and source their own food. Until recently such practices have been more common in the developing world, perhaps most famously the urban gardens of Havana, Cuba, which produce over 90 per cent of the city's food (Koont, 2009, based on figures for 2002). In the developed world the increasing interest in self-sufficiency generated by movements such as the 'Transition

Towns' network has prompted a resurgence in urban gardening, for example a highly successful project in Todmorden, UK (IET, 2011).

3.8 Human behaviour

One of the greatest remaining challenges for reducing emissions from all economic sectors is understanding, modelling and influencing human behaviour. In an ideal world it would be sufficient to provide information on how to reduce energy and resource use and let rational decision making do the rest. However, human society is complex, and humans often behave far from rationally – or conversely it may be completely rational for someone not to adopt a 'pro-environmental' behaviour if the costs and implications of doing so are deemed to be unacceptable, perhaps for financial reasons or any additional time required. Furthermore, humans are far from united in their attitudes towards climate change, and therefore enabling effective behaviour change requires understanding and addressing these differences (Scottish Government, 2009b). Another challenge here is how to reinforce such behaviours so they become habits rather than conscious actions, for example remembering to switch off unused lights and appliances (Jackson, 2005).

Human attitudes towards the environment, and particularly towards climate change, mean that even simple measures to influence behaviour can have unintended outcomes. Some of these outcomes may be deliberate; for example, when WWF launched it's 'unprintable' .wwf file format the reaction on the internet was not entirely positive, with some people posting workarounds and others stating their intentions to behave in the opposite way to that intended (Carrington, 2011). Nevertheless, the built environment presents many opportunities to reduce energy and resource consumption by influencing human behaviour that do not depend on attitudes or awareness. Some examples of these include the following:

* installing programmable controls on space and water heating systems that allow users greater ability to configure them to their needs;
* fitting timer switches or motion sensors to lighting systems, and ensuring they are set to operate for the most optimal period of time;
* fitting spray taps and setting water heating to the optimal temperature;
* providing recycling bins close to where waste is produced, and limiting the availability of bins for general waste;
* installing software that automatically powers down computers after periods of inactivity, and that cannot be easily circumvented by users;
* providing more and better public transport, cycling infrastructure, and open or green spaces in urban areas (Baker, 2007).

However, even the most subtle and useful attempts to influence behaviour may not go according to plan – see Chapter 6 for an example of one of these.

Bibliography

All websites were last accessed on 30 November 2011.

Ainger, C., Baker, K., Crishna, N., Cziganyik, N., Fenn, T., Johnson, A., Jowitt, P., Robertson, C. and Smith, C. 2008. *Carbon Accounting in the UK Water Industry: Guidelines for Dealing with 'Embodied Carbon' and Whole Life Carbon Accounting*. London: UK Water Industry Research.

Allen, S.R., Hammond, G.P. and McManus, M.C. 2008. Prospects for and barriers to domestic micro-generation: a United Kingdom perspective. *Applied Energy*, **85**, pp. 528–544.

Baker, K.J. 2007. Sustainable cities: determining indicators of domestic energy consumption. Ph.D. thesis, Institute of Energy and Sustainable Development (IESD), De Montfort University, Leicester.

Bannister, D. 2008. The sustainable mobility paradigm. *Transport Policy*, **15**, pp. 73–80.

Bennett, A.F. 2003. *Linkages in the Landscape: The Role of Corridors and Connectivity in Wildlife Conservation*, IUCN Forest Conservation Programme. Gland: IUCN. Available at: http://www.app.icun.org/dbtw-wpd/edocs/FR-021.pdf.

Blakey, S., Rye, L. and Wilson, C.W. 2011. Aviation gas turbine alternative fuels: a review. *Proceedings of the Combustion Institute*, **33**, pp. 2863–2885.

Blalock, T. 2006. Powering the New Yorker: a hotel's unique direct current system. *IEEE Power and Energy Magazine*, January/February.

Brito Cruz, C.H. de. 2008. *Bioethanol in Brazil*. São Paulo: Fapesp (São Paulo Research Foundation). Available at: http://www.biofuels.apec.org/pdfs/apec_200810_brito-cruz.pdf.

Bruvoll, A. and Fæhn, T. 2006. Transboundary effects of environmental policy: markets and emission leakages. *Ecological Economics*, **59**, pp. 499–510.

Carrington, D. 2011. WWF launches PDF-like file format that can't be printed. *Guardian*, 6 February. Available at: http://www.guardian.co.uk/environment/damian-carrington-blog/2011/feb/09/wwf-pdf-file-format-printed.

Ciotola, R.J., Lansing, S. and Martin, J.F. 2011. Energy analysis of biogas production and electricity generation from small-scale agricultural digesters. *Ecological Engineering*, **37**, pp. 1681–1691.

CST. 2009. *Improving Innovation in the Water Industry: 21st Century Challenges and Opportunities*, March. London: UK Council for Science and Technology. Available at: http://www.bis.gov.uk/assets/bispartners/cst/docs/files/whats-new/09-1632-improving-innovation-water-industry.pdf.

DECC. 2009. *UK Low Carbon Transition Plan*. London: UK Department for Energy and Climate Change. Available at: http://www.decc.gov.uk/assets/decc/White%20Papers/UK%20Low%20Carbon%20Transition%20Plan%20WP09/1_20090724153238_e_@@_lowcarbontransitionplan.pdf.

DECC. 2011. *Digest of UK Energy Statistics (DUKES), 2011*. London: Stationery Office. Available at: http://www.decc.gov.uk/en/content/cms/statistics/publications/dukes/dukes.aspx.

De Paepe, M., D'Herdt, P. and Mertens, D. 2006. Micro-CHP systems for residential applications. *Energy Conversion and Management*, **47**, pp. 3435–3446.

EA. 2009. *A Low Carbon Water Industry in 2050*. London: UK Environment Agency. Available at: http://publications.environment-agency.gov.uk/PDF/SCHO1209BROB-E-E.pdf.

Emery, A., Davies, A., Griffiths, A. and Williams, K. 2007. Environmental and economic modelling: a case study of municipal solid waste management scenarios in Wales. *Resources, Conservation and Recycling*, **49**, pp. 244–263.

Envac. 2009. *Vacuum Technology*. Stockholm: Envac AB. Available at: http://www.envacgroup.com/web/Vacuum_technology.aspx.

ETN – Emerging Technologies News. 2011. Solar sail technology the powering force behind international shipping, 28 February. Available at: http://www.emerging-technologies-news.info/index.php/new-technology/solar-sail-technology-powering-international-shipping/.

EU Directive 2005/32/EC. Known as the 'Energy Using Products Directive'. Full title: 'Establishing a framework for the setting of ecodesign requirements for energy-using products and amending Council Directive 92/42/EEC and Directives 96/57/EC and 2000/55/EC of the European Parliament and of the Council'.

EU Directive 2009/125/EC. Known as the 'Ecodesign Directive'. Full title: 'Directive 2009/125/EC of the European Parliament and of the Council of 21 October 2009 establishing a framework for the setting of ecodesign requirements for energy-related products'.

Gellynck, X., Jacobsen, R. and Verhelst, P. 2011. Identifying the key factors in increasing recycling and reducing residual household waste: a case study of the Flemish region of Belgium. *Journal of Environmental Management*, **92**, pp. 2683–2690.

Hammond, G. and Jones, C. 2011. *Inventory of Carbon and Energy*. Bath: University of Bath. Available at: http://www.bath.ac.uk/mech-eng/sert/embodied/.

Helm, D., Smale, R. and Phillips, J. 2007. Too good to be true? The UK's climate change record. Available at: http://www.dieterhelm.co.uk/sites/default/files/Carbon_record_2007.pdf.

Houston, K.J. 2011. The impact of ICT on energy consumption and GHG emissions. Available at: http://www.icarb.org/presentations.

IET. 2011. *Incredible Edible Todmorden*. Todmorden: IET. Available at: http://www.incredible-edible-todmorden.co.uk.

Jackson, T. 2005. *SDRN Briefing Note One: Motivating Sustainable Consumption*. London: Sustainable Development Research Network. Available at: http://www.sd-research.org.uk/wp-content/uploads/sdrnbriefing1motivatingsustainableconsumption_001.pdf.

Koont, S. 2009. The urban agriculture of Havana. *Monthly Review*, **60**. Available at: http://www.monthlyreview.org/2009/01/01/the-urban-agriculture-of-havana.

Lee, J. 2007. Off goes the power current started by Thomas Edison. *New York Times*, 14 November. Available at: http://cityroom.blogs.nytimes.com/2007/11/14/off-goes-the-power-current-started-by-thomas-edison/.

Marshall, S. 2001. The challenge of sustainable transport. In *Planning for a Sustainable Future*, ed. A. Layard, S. Davoudi and S. Batty, pp. 131–147. London: Spon.

Martin, M., Williams, I.D. and Clark, M. 2006. Social, cultural and structural influences on household waste recycling: a case study. *Resources, Conservation and Recycling*, **48**, pp. 357–395.

McNichol, T. 2006. *AC/DC: The Savage Tale of the First Standards War*. New York: John Wiley & Sons.

Mercy Corps. 2009. *Bosnia: Biodiesel from Recycled Cooking Oil*, June. Edinburgh: Mercy Corps. Available at: http://www.mercycorps.org.uk/countries/bosniaandherzegovina/11785.

Perrin, D. and Barton, J. 2001. Issues associated with transforming household attitudes and opinions into materials recovery: a review of two kerbside recycling schemes. *Resources, Conservation and Recycling*, **33**, pp. 61–74.

Resch, G., Held, A., Faber, T., Panzer, C., Toro, F. and Haas, R. 2008. Potentials and prospects for renewable energies at global scale. *Energy Policy*, **36**, pp. 4048–4056.

Scottish Government. 2009a. *The Scottish Soil Framework*. Edinburgh: Scottish Government. Available at: http://www.scotland.gov.uk/Publications/2009/05/20145602/0.

Scottish Government. 2009b. *SEABS 08: The Scottish Environmental Attitudes and Behaviours Survey 2008*. Edinburgh: Scottish Government Social Research. Available at: http://www.scotland.gov.uk/Publications/2009/08/03100422/0.

SEPA. 2011. *The Waste Hierarchy*. Stirling: Scottish Environmental Protection Agency. Available at: http://www.sepa.org.uk/waste/moving_towards_zero_waste/waste_hierarchy.aspx.

Shukla, A., Pekny, J. and Venkatasubramanian, V. 2011. An optimization framework for cost effective design of refueling station infrastructure for alternative fuel vehicles. *Computers and Chemical Engineering*, **35**, pp. 1431–1438.

Sidique, S.F., Lupi, F. and Joshi, S.V. 2010. The effects of behavior and attitudes on drop-off recycling activities. *Resources, Conservation and Recycling*, **54**, pp. 163–170.

Stromberg, J. 2004. Stockholm goes for biogas buses. Available at: http://www.trendsetter-europe.org/index.php?ID=2842.

Swider, D.J., Beurskens, L., Davidson, S., Twidell, J., Pyrko, J., Prüggler, W., Auer, H., Vertin, K. and Skema, R. 2008. Conditions and costs for renewables electricity grid connection: examples in Europe. *Renewable Energy*, **33**, pp. 1832–1842.

Tiwari, R., Cervero, R. and Schipper, L. 2011. Driving CO_2 reduction by integrating transport and urban design strategies. *Cities*, **28**, pp. 394–405.

Torekov, M.S., Bahnsen, N. and Qvale, B. 2007. The relative competitive positions of the alternative means for domestic heating. *Energy*, **32**, pp. 627–633.

Townsend-Small, A. and Czimczik, C.I. 2010. Carbon sequestration and greenhouse gas emissions in urban turf. *Geophysical Research Letters*, **37**, L02707.

Water UK. 2009. *How the Water Industry Is Managing Its Contribution to Climate Change*. London: Water UK. Available at: http://www.water.org.uk.

Wissner, M. 2010. Smart grids – a saucerful of secrets? *Applied Energy*, **88**, pp. 2509–2518.

Section B

Strategies for a low carbon built environment

4 Energy generation for a low carbon built environment

Determining which technologies to invest in to best meet our increasing energy demands whilst decarbonising energy supply is probably the most divisive topic for those working in the fields of energy and the built environment. Very few would argue with the need to move away from fossil fuels, but towards what, and how fast? Experts agree that the 'what' needs to be a mix of technologies, but in what proportions, and are there any that should be excluded?

Humanity has yet to develop a low carbon energy technology that can meet demand whilst being acceptable to everyone, and many alternative technologies face significant public opposition. Even writing about some of these immediately leads to accusations of bias in favour of whichever technologies fare better in the analyses.

The first summaries presented here focus on micro and distributed generation technologies that are currently on the market. Where the efficiency of a technology is expected to improve without significant innovation, this is noted. The costs quoted for each technology come from a range of sources (primarily from the UK and the USA) and exclude savings from subsidies. They are intended to represent reasonable boundaries but should not be considered absolutes.

Owing to the range of financial incentives for different technologies that are available around the world and the frequencies with which they change, any summary of these would be extensive and quickly out of date. However, understanding the types and levels of incentives that apply to each technology, as well as any relevant local, national and international legislation and regulation, is essential in selecting the most appropriate technology (or technologies) for each individual need.

Geography also has a huge role to play, both local and global. In the case of renewables, solar is obviously most efficient at lower latitudes, whereas wind and wave tend to favour more exposed locations at higher latitudes. However, solar is perfectly viable even at high latitudes, and exploiting manufactured wind tunnels in the built environment may provide new sources of wind power. Similarly the viability of combined heat and power (CHP) depends significantly on the local built environment, but there also needs to be a reliable fuel supply; and the selection of any large scale technology has to consider the local availability of the resource and the carbon cost (and other impacts) of any fuel imports.

Nevertheless, this chapter is intended to provide a brief overview of the current renewable and low carbon alternatives to fossil fuels, with a particular focus on micro and distributed generation in the built environment.

4.1 Micro and distributed generation

Micro and distributed energy generation technologies are an increasingly common feature of built environments and are expected to have a significant role to play in shaping our energy future. Some of these technologies are more visible than others; for example, wind turbines installed on rooftops stand out more than solar panels (which may go almost unnoticed from below when installed on urban infrastructure), whereas ground source heat pumps are largely invisible after installation. Some are more flexible than others; for example, solar lends itself to a huge range of applications, whereas micro wind and hydro are more resource dependent. However, inevitably, cost is a major factor in determining the appropriateness of each technology for any given application, and critically so is the 'payback period' – the length of time needed to recoup the investment from generating electricity and/or heat. Therefore, given the cost per unit output from these technologies, the availability of subsidies and other financial incentives, along with the long term financial signals (i.e. political and corporate commitments that any incentives will not be open to sudden and drastic changes), is essential in building investor confidence – whether those investors are individual householders or large organisations.

However, in most countries the key barrier to enabling greater uptake is electricity infrastructure designed for high output, centralised, fossil fuelled generation (see 3.1, 'Energy generation'). The costs of restructuring inevitably lead some to argue that strategies to decarbonise electricity and heat supplies should focus on centralised renewable generation, particularly higher output options such as wind and solar farms. Yet these tend to be most suited to locations away from large human settlements and therefore also require significant investment in new infrastructure.

In reality the future of renewable and low carbon energy generation looks set to be a healthy mix of types, scales and applications of different technologies. Centralised renewables and low carbon technologies will be essential in ensuring 'baseload power' (the minimum level needed to ensure that heavy industry and essential services can operate uninterrupted at all times). However, the significant potential of micro-generation technologies to harness local resources to meet energy demands, particularly for off-grid properties and those in highly urbanised areas, makes them an essential tool for reducing emissions from the built environment, whilst at the intermediate scale 'community level' applications of distributed generation technologies look set to grow in their contribution to meeting energy demands. These include stand-alone wind turbines, ground source heat pumps, and CHP.

Table 4.1 provides an overview of the most common micro and distributed generation technologies appropriate for urban and rural built environments.

Table 4.1 Comparative advantages and disadvantages of different distributed and micro-generation technologies

Technology	Typical costs (UK)	Advantages	Disadvantages
Solar thermal	£2,000–£4,500	Solar resource is relatively reliable and predictable. Proven technology. Low cost (compared to other renewables). Systems can also be powered renewably (e.g. solar PV powered pump). Proven/established technology. Visually unobtrusive. Provides hot water all year round (however, will not meet demand in winter). Low maintenance. Significant potential capacity (unused roof spaces).	Some systems require grid electricity supply. Low cost reduction potential due to established designs. Does not generate electricity.
Solar PV	£6,000–£15,000	The solar resource is relatively reliable and predictable. Proven and advancing technology, with significant potential for further cost reduction. Visually unobtrusive. Low maintenance. Significant potential capacity (unused roof spaces).	Relatively high capital costs. Not always cost-effective without subsidies or incentives.
Micro wind	£3,000–£5,000	Proven technology, with some potential for increased efficiency. Can be relatively inexpensive when situated appropriately. Matches loosely with daily variations in energy demand. Wide scale of installations/outputs available.	Very site-specific resource in urban areas. Least predictable intermittent renewable. Lack of available performance information. Some opposition (e.g. because of visual impact). May be subject to local building or planning regulations (e.g. because of noise and vibration issues).
Ground source heat pump	£8,000–£17,000	Very reliable – ground temperatures are constant and predictable. Can be cost-effective within the current market.	Retrofitting can be problematic (most effective with under-floor heating). Requires relatively large electricity supply. Land requirement for

Table 4.1 Continued

Technology	Typical costs (UK)	Advantages	Disadvantages
			ground loops. Limited applicability in densely populated areas due to impact on ground temperatures. High capital costs. Installation as retrofit requires significant disruption.
Air source heat pumps	£5,000–£10,000	Can provide hot air or water for heating. Air is a free and unlimited resource. Widely applicable, including as retrofits, so high potential for emissions reduction. Efficiencies now improving, but generally unsuitable for cold climates. Units may be visually obtrusive when retrofitted (similar to air conditioning units).	Does not generate electricity. Requires an electricity or gas supply. (but this can be powered by renewables)
Geothermal	£3,000–£6,000 for small scale (estimated from US figures and excluding significant non-build costs)	Provides an almost uninterrupted supply (higher than for even fossil fuel plants). Low maintenance and low cost of operation. Long term sustainability of extracting heat from geothermal hot spots has been demonstrated.	Highly location-dependent resource. Significant costs incurred for identifying new geothermal hot spots.
Micro and community CHP and biomass	Micro: approximately £3,000 Community: varies widely according to scale and infrastructure costs	Has the potential to reduce CO_2 emissions related to fossil fuel use through efficiency gains. Technologies at or nearing cost-effectiveness under current market conditions. Biomass can be used as a fuel source. Community scale applications can provide low (or no) cost heating to large urban areas. Relatively short payback periods, even at community scale.	CHP units currently commonly fossil fuel powered. Has an inflexible heat to power generation ratio, which can be problematic if this does not match the respective demands. Noise levels for domestic units may be unsuitable for some small homes or flats. Carbon savings appear to be less than originally predicted. Some localised pollution concerns. Lack of available performance information. Use of biomass requires fuel to be available locally. Land area required for growing biomass fuel is a cause for concern. Community scale application limited by the availability of suitable locations.

Anaerobic digestion	£100,000–£2 million: for example, £150,000 for a 25kWe agricultural waste AD plant – but varies by scale and waste sources	Utilises waste to generate heat and/or electricity. May divert waste from landfills. Solid and liquid by-products can be used as fertiliser. Integrates waste management from a wide range of sources (domestic, agriculture, etc.) with energy generation (and CHP). High costs offset by high revenues.	Requires a sufficient and reliable waste stream. Quality (and toxicity) of by-products depends on quality of waste. May divert waste from recycling streams. May cause local odour problems.
Micro-hydro	Average £25,000 for a 5kW system, but highly variable	High energy yields possible. Proven technology. High potential for expansion in many countries. Reliable resource, but varies seasonally. Wide scale of installations/outputs available. Can be cost-effective within the current market, even given high capital costs.	Site-specific resource. Application limited by the availability of suitable locations. May be subject to conservation legislation.

Sources: Allen, Hammond and McManus, 2008; Bahaj, Myers and James, 2007; BEC, 2011; CHPA, 2011a; DECC, 2011a, 2011b, 2011c; DFPNI, 2010; EST, 2005, 2011a, 2011b; GEA, 2011; Kutscher, 2001; MacKay, 2011; Mulliner, 2011; REN, 2011; Scottish Government, 2008; Suzuki *et al.*, 2009; US DoE, 2006, 2011a, 2011b; Weber and Shah, 2011.

4.1.1 Solar thermal

Solar thermal or solar hot water (SHW) panels are probably the most commonly installed building-integrated renewable technology. Solar thermal panels (also known as 'collectors') can be fitted at optimal angles on rooftops and contain a liquid, usually an antifreeze, which is heated by the sun and pumped to heat water in a boiler. Although output is dependent on weather conditions, the technology can provide hot water all year round, even in higher latitudes. For example, in temperate countries such as the UK building-mounted solar thermal panels currently have the potential to meet up to 70 per cent of an average household's hot water needs, with the additional benefits of being low maintenance and low cost in comparison to other micro renewables (DECC, 2011a).

The availability of used roof space in urban areas means that both solar thermal and solar photovoltaics have significant potential to reduce carbon emissions from the built environment. However, in the short term solar thermal may have a particularly important role to play in enabling a shift away from using natural gas to meeting heating demands.

4.1.2 Solar photovoltaics

Solar photovoltaic cells (PVs) convert energy from the sun directly into electricity, and are a proven and highly popular renewable technology that is still rapidly advancing. As for solar thermal systems, PV panels can be installed on any roof with an appropriate aspect, but they can also be integrated into roofing tiles, and walls, and new thin film designs can be affixed to windows. One of the most common places to find a panel is on top of transport infrastructure such as parking meters, and trials of PVs integrated into road surfaces are being conducted in Oregon, USA (OIPAF, 2008).

Historically PV has suffered from a lack of investment, and this helps explain why costs remain a barrier to wider installation, as high costs mean long payback periods for investors. As the efficiency of solar cells continues to improve and the costs of manufacture continue to fall, along with an increasing range of applications in the built environment (including powering other renewables), PV is set to be one of the most important technologies for reducing GHG emissions from the built environment.

4.1.3 Micro wind

Wind turbines come in a wide range of designs and sizes, which maximises their ability to generate electricity from any available wind resource. Micro wind includes both building-mounted turbines, typically capable of generating anything up to 2kW, and the smaller stand-alone turbines commonly used by off-grid buildings. Most designs are horizontally mounted, and many share the three-blade design used for many larger turbines, but various blade configurations are available and vertically mounted turbines are suitable for smaller stand-alone installations.

A key problem for micro wind in urban environments is that the complexity and variation in local air flows can result in higher intermittency in supply than for other micro renewables (Weber and Shah, 2011). Noise and vibration may also pose problems for mounting turbines on existing buildings, and so as with any micro renewable they may be subject to local planning laws. Nevertheless, the flexibility of micro wind turbines makes them another valuable option for reducing emissions.

4.1.4 Ground source heat pumps

Ground source heat pumps (GSHPs) use heat pumps to utilise the stable temperature of the ground to provide heating and/or cooling for both space and water. They are distinct from geothermal systems in that they are not limited by the need to identify and exploit geothermal 'hot spots' (the heat comes from the sun, not the earth), and the thermal stability of the ground makes them more efficient than their air source equivalents. Heat (or cooling) is delivered by pumping a fluid with a high thermal capacity and low freezing point around a 'loop' installed below ground and through a heat exchanger on the surface. Although they require electricity for powering the pump, this can be delivered by solar PV (creating 'geo-solar' systems) and, once installed, they are low maintenance and have long lifespans, typically 25 years for the pump and 50 or more for the loop, and they also offer lower payback periods than some other renewables (US DoE, 2011b).

However, GSHPs are not without their disadvantages, particularly for applications in urban areas. Although loops can be installed under existing buildings, this entails significant disruption from construction; for example, installing one under a domestic property may require digging up any garden area, and higher concentrations of GSHPs can change ground temperatures, leading to reduced system efficiencies. GSHPs are also not suitable for use in the colder climates found towards the poles (MacKay, 2011).

Although less flexible in application than most other renewables, GSHPs provide consistent and long term supplies of heat and cooling, and when combined with solar PV provide an important source of zero carbon energy generation.

4.1.5 Air source heat pumps

Air source heat pumps (ASHPs) are a relatively new addition to the options for harnessing the renewable energy potential of built environments. From the outside they resemble, and may be mistaken for, the air conditioning units installed on buildings in cities around the world, and the relative simplicity of retrofitting ASHPs is one of their key advantages.

ASHPs operate on the same principles as GSHPs, but use air as a heat exchanger instead of the ground. The use of air, with its lower thermal capacity and much higher temperature instability, means that ASHPs are less efficient than GSHPs (GSHPs being more than twice as efficient on cold days in temperate climates), although performance is improving. Also like GSHPs they require an electricity supply that can be met from solar PV, and in suitable locations they can also be combined with GSHPs for greater efficiencies at lower marginal costs (US DoE, 2011b).

The greater flexibility of ASHPs, particularly for applications in densely populated areas and on high rise buildings, means that they are expected to become an increasingly common sight in urban environments.

4.1.6 Geothermal

Geothermal is included here mainly for the purposes of distinguishing it from ground source heat pumps. The term covers a range of technologies that utilise heat in geothermal 'hot spots' in the earth's crust to generate heat and electricity by pumping water through them, which in more modern designs is recycled to limit resource consumption and help reduce operational emissions to near zero. The limited availability of hot spots and the costs incurred from locating them mean that geothermal plants remain a fringe technology at the small scale. However, small plant construction costs compare favourably with other renewables and are highly cost-effective once in operation.

Geothermal may be a much more attractive option at the large scale. Even though set-up costs (including exploration) dwarf those of conventional fossil fuel plants, they suffer from a much lower rate of supply interruption – average availabilities for geothermal being around 90 per cent, compared to around 75 per cent for coal plants – and they can provide an almost limitless output of cheap low carbon energy (GEA, 2011; Kutscher, 2001; US DoE, 2006).

4.1.7 Micro and community CHP and biomass

All combined heat and power technologies share the common characteristic of burning a fuel (natural gas or biomass) to generate electricity whilst capturing and utilising the heat that would otherwise be wasted. At the micro-scale they can be retrofitted in place of conventional boilers, and the dimensions of most common models are designed to facilitate this. Power plants for community scale systems (also known as district heating) can be housed in public or commercial buildings, for example in council offices such as Leicester City Council in the UK (CHPA, 2011b) and leisure centres such as Ards, near Belfast, in Northern Ireland (DFPNI, 2010). Both of these schemes are retrofits to existing urban environments. However, community CHP can be even more effective when designed into new urban developments, such as Hammarby Sjöstad in Sweden (Suzuki *et al.*, 2009).

CHP is not without its criticisms, predominantly over fuel use, which is invariably natural gas or biomass. Using natural gas means reliance on a fossil fuel and its long term costs and security of supply, whereas using biomass requires access to a regular fuel supply that carries significant concerns over its sustainability and impacts on climate change (see 4.2.5 below). Nevertheless CHP, in all its forms, is a very attractive option for decarbonising urban environments for a range of reasons. First and foremost is cost-effectiveness at the large scales – both UK examples listed here have payback periods of under five years – which makes CHP particularly suitable for tackling urban deprivation. Another advantage, assuming the availability of a regular fuel supply, is its ability to meet baseload demand (in contrast to many other renewables). Higher power outputs reduce the technical difficulties in connecting to existing electricity infrastructure, gas fuelled plants can be converted or replaced to use biomass, and, for those with an eye further to the future, CHP is a 'transition technology' that requires infrastructure that could one day be upgraded for fuel cells. In the immediate term a major role for CHP will be facilitating the shift away from using natural gas for heating whilst the capacity of renewable electricity and heating technologies scales up to meet demand (Weber and Shah, 2011).

4.1.8 Anaerobic digestion

Anaerobic digestion (AD) is the process whereby bacteria break down organic material in the absence of air, yielding a biogas containing methane which can be burnt to generate electricity and heat. Although generally associated with agricultural waste and sewage, which provide more consistent fuel than mixed domestic and commercial waste, it is now expanding into more urban areas. AD offers potentially significant system and cost-efficiency benefits through integrating energy generation with waste infrastructure, as well as being capable of producing fertilisers as by-products, and can be integrated into existing CHP networks. This means that, whilst the costs of AD plants are at the high end of those for distributed generation technologies, the revenues that can be gained from the sale of energy and fertiliser are significant enough to make AD a commercially viable technology (BEC, 2011; Mulliner, 2011).

AD is not without its critics, not least of whom will be those living near plants that do not adequately control their odour emissions – hence the progress with agricultural waste and sewage-fed plants that are sited in locations with existing odour problems. The mixed waste plants used for domestic and commercial waste have also been accused of diverting waste from entering recycling streams. AD is also the latest evolution of biogas plant technologies, some of which (most famously the low tech Chinese digesters) have entered popular consciousness by exploding. Although modern AD plants are safe, and highly regulated in many countries, these problems can make all waste to energy conversion plants (often collectively but erroneously termed 'incinerators') publicly unpopular.

Nevertheless the versatility of AD plants, especially when integrated with CHP networks, means that they can be expected to make an increasing contribution to the supply of low carbon heat and electricity. For more on AD see 3.4, 'Waste management'.

4.1.9 Micro-hydro

At first glance it may seem strange to include micro-hydro in a list of technologies appropriate to urban environments, as it tends to conjure up images of rural idylls surrounded by mountains and streams. However, it is important to remember that many major cities still contain waterways that may be ideally suited to micro-hydro installations, and even rural areas encompass smaller built environments that contribute to national energy demands. A telling statistic from the UK is that although only 56 micro-hydro installations were in operation by 2011 their combined output was nearly 3TWh, which was more than for either solar PV or wind, despite micro-hydro installations being far fewer in number (DECC, 2011c).

Most micro-hydro installations are 'run of the river', meaning that they do not use dams to build up a reservoir of water but simply divert part of the river or stream through one or more turbines, which means that they can be installed without the problems associated with damming rivers and flooding land upstream. Their output is dependent on both the 'head' (the height from which water falls into the turbine) and the volume and rate of water flowing through them, and so they come in a wide range of scales, designs and outputs – often subdivided further than the simple micro versus large scale convention used here.

The potential of micro-hydro is limited by comparatively fewer barriers than for other micro renewables, in particular cost-effectiveness, although ecological considerations such as breeding grounds for river fauna may exclude their installation on some rivers or require additional mitigation measures such as air curtains to deter fish from entering the turbines from downstream. In light of these advantages it is no surprise that many countries, such as Scotland, are actively identifying and utilising their micro-hydro resources (Scottish Government, 2008).

4.2 Centralised renewable generation

As the focus of this book is on the built environment it is beyond its scope to provide a detailed description of large scale centralised renewable and low carbon technology. Therefore the intention here is simply to provide an overview of the key currently available technologies, and highlight some of the most important issues that surround them.

The urgent need to move away from, and ultimately cease, our dependency on fossil fuels means that all these technologies will have some role to play in meeting emissions reduction targets, and rapid progress around the globe meant that for 2010 renewables accounted for almost half of newly installed electricity capacity (REN, 2011).

4.2.1 Hydropower

Hydroelectric dams are humanity's great monuments to the early days of renewable energy, although originally motivated by the need to generate large amounts of power without the need to transport fuel rather than for their emissions credentials. The controversies that surround hydropower dams usually relate to their impacts on landscapes, local human and animal populations, and the flow and quality of water downstream – the latter being particularly controversial when a river crosses state or national borders. An excellent case study of these debates can be gleaned from the volumes of work published on the Colorado River Compact in the USA. From an emissions perspective it is debatable whether hydropower at this scale is 100 per cent renewable, as flooding land produces significant amounts of emissions, particularly methane. Hydropower is also highly location-dependent, and new dams are often subject to a wide range of legislation and other limiting factors, with some countries already having exploited much of their available potential.

Table 4.2 Global generation from renewables, 2011

Technology	World total	EU-27	USA	China	India	Developing countries
Wind	198	84	40	45	13	61
Biomass	62	20	10	4	3	27
Solar PV	40	29	2.5	0.9	~0	n/a
Geothermal	11	1	3.1	~0	0	5
Solar thermal	1.1	0.6	0.5	0	0	0
Wave/tidal	0.3	0.3	0	0	0	0
Total renewable capacity (excluding hydropower)	312	135	56	50	26	94
Hydropower	1,010[1]	130	78[2]	213	40[2]	n/a
Total renewable capacity (including hydropower)	1,320	265	134	263	56	n/a

Source: REN, 2011.

Notes:

All figures exclude installations below several MW, and therefore do not reflect micro-generation capacity.
Owing to inconsistencies in source data, hydropower includes some pumped storage, except where noted.
1 Rounded to nearest 10GW.
2 Conventional hydropower only.

4.2.2 Wind farms

After hydro, by far the most widely installed centralised renewable is wind power, either on-shore or off-shore. The tendency for the on-shore resource potential to be greatest in exposed and picturesque rural areas has generated significant public opposition in some parts of the world, but those in favour of greater expansion argue that the immediate visual impacts are far outweighed by the long term impacts of climate change. Off-shore wind farms use larger turbines and produce much higher outputs of electricity. However, the difficulties of constructing farms far out to sea mean that at present most are still visible from the shore. One solution to this, as used in countries such as Germany, is to locate them alongside existing transport networks. Both on-shore and off-shore produce intermittent supplies of electricity and usually require new electricity infrastructure. Nevertheless, they form an essential component of the collection of renewables that can meet existing energy demands using proven and commercially viable technologies.

4.2.3 Solar farms

When installed at large scales both solar thermal and solar PVs can be used to generate electricity. Solar farms consist of either large arrays of PV panels or vast thermal plants that use mirrors to focus energy from the sun on to a heat transfer fluid. Most commonly the latter is achieved by using parabolic mirrors to focus energy on a tube containing the fluid, but more recent designs contain it in a tower surrounded by a circular array of mirrors that focus the energy on its tip. Such installations are also termed concentrating solar power (CSP) farms, and the same principles can be applied to improve the output from PVs, in what are termed concentrating photovoltaic (CPV) farms.

Although some CSP farms have existed for many years, particularly in the USA, it is only recently that the technology has really taken off. Farms now being developed will significantly ramp up global capacity from around just over 1.1GW to more than 17GW. Almost half of this expansion is in the USA (8.7GW), followed by Spain (4.5GW) and China (2.5GW). Although solar farms are relatively uncommon at present, both forms of the technology are expected to play an increasingly significant role in reducing emissions and are a cornerstone of major infrastructure projects such as the European Supergrid (REN, 2011; Wang, 2011).

4.2.4 Wave and tidal power

Wave and tidal power covers a wide and diverse range of technologies, which can be split into two broad categories according to whether they use static turbines or other technologies. The use of static turbines is a proven and generally commercially viable option that includes installing them in river barrages to generate electricity from the river flow, or along shorelines to capture energy from incoming and outgoing waves. Newer designs may be installed on the sea bed. All of these come with their environmental and ecological impacts, which can generate significant public opposition, such as that against the various proposals for a barrage on the River Severn in the UK, which has the second highest tidal range in the world.

The alternatives are a rapidly evolving mix of technologies which can trace their roots to 'Salter's Duck', developed by the Scottish scientist Stephen Salter in response to the oil crisis of the 1970s. The device, also known as a 'nodding duck', generates electricity from gyroscopes housed in a wedge-shaped case that floats on the surface of the water (Moss, 1982). In more recent years the range of these technologies has undergone a rapid expansion, and their scale has increased significantly. Probably the most well-known wave power device today is the Pelamis™ (or 'sea snake'), which was also invented in Scotland.

4.2.5 *Biomass*

Using biomass to generate electricity means setting aside areas of land from which to source the fuel. This means that, whilst small to medium scale biomass has some potential to provide a sustainable source of low carbon energy, larger scale applications of the technology can be highly controversial. Sourcing (invariably importing) the volume of biomass needed to supply large scale biomass plants is a very different proposition from sourcing biomass for smaller scale installations fed from local, sustainable and regulated producers. The same problem applies to converting transport to run on biofuels (see 3.2, 'Transport'). Although biomass is generally classified as a renewable, the emissions from transporting fuel for large scale biomass and biofuel projects arguably define them as low carbon technologies.

4.2.6 *Nuclear*

Nuclear power remains by far the most divisive low carbon technology, to the point that it is impossible to publish anything on the subject without facing allegations of bias from either of the firmly entrenched camps of supporters and opponents. It is probably safe to state that nuclear power is unquestionably a technologically and commercially viable means of providing large and uninterrupted amounts of low carbon electricity from centralised plants, and therefore cannot be discounted as an alternative to fossil fuels. A summary of the most important advantages and disadvantages is given in Table 4.3.

The most prominent concern over nuclear power relates to its wider impacts and so will not be addressed here. However, from the perspective of reducing emissions from the built environment it is important to consider the extent to which global emissions reduction strategies rely on

Table 4.3 Advantages and disadvantages of nuclear power

Advantages	Disadvantages
Large output – can meet baseload demand from a small number of installations. Uninterrupted supply. Small volume of fuel required (so mining impacts lower than for fossil fuels). Centralised technology, so generally suited to current grid structures. Proven technology, still advancing.	Safety concerns following high profile disasters. Produces nuclear waste. Older designs produce military-grade waste. Costs of development, operation, decommissioning and long term storage of waste. Contested, but large, embodied energy. Significant public and political opposition.

Table 4.4 Top ten per capita generators of nuclear energy, 2007

Country	kWh/d per capita
Sweden	19.6
France	19.0
Belgium	12.2
Finland	11.8
Switzerland	9.7
South Korea	7.7
USA	7.5
Canada	7.4
Slovenia	7.4
Slovakia	7.2

Source: MacKay, 2011.

grid decarbonisation. Although contributions from true renewables are rising rapidly, global energy demand is still increasing, and as yet there is no sign that renewables will be able to meet this demand completely within the time available for meeting global emissions reduction targets.

Nuclear is also a major source of debate for carbon accountants, as the resolving of uncertainties and arguments around the carbon embodied in nuclear plants and their whole life costs will be important evidence in determining their future.

In many countries the main alternatives to meeting the current and immediate future supply–demand gap that are being proposed are coal and gas plants equipped with carbon capture and storage technology (CCS), coal gasification (CG) and large scale biomass (see 4.2.5 above). Whilst the volume and sustainability of fuel supplies seriously limit the potential of biomass, coal remains relatively abundant and a tempting source of cheap energy. Both CCS and CG have yet to be proven either technologically or commercially viable at a large scale, both use fossil fuels, and both have serious environmental and safety concerns.

Making the right decisions over our global energy future requires gathering and understanding the best evidence researchers can provide, and carbon accounting can make significant contributions to the debates. We will return to these discussions in Chapters 11 and 12.

References

All websites were last accessed on 30 November 2011.

Allen, S.R., Hammond, G.P. and McManus, M.C. 2008. Prospects for and barriers to domestic micro-generation: a United Kingdom perspective. *Applied Energy*, **85**, pp. 528–544.

Bahaj, A.S., Myers, L. and James, P.A.B. 2007. Urban energy generation: influence of micro-wind turbine output on electricity consumption in buildings. *Energy and Buildings*, **39**, pp. 154–165.

BEC, 2011. *Anaerobic Digestion*. Farnham: Biomass Energy Centre. Available at: www.biomassenergycentre. org.uk/portal/page?_pageid=75,17509&_dad=portal&_schema=PORTAL.

CHPA. 2011a. *Guide to Community Heating and CHP: Commercial, Public and Domestic Applications*. London: Community Heat and Power Association (UK). Available at: www.chpa.co.uk/medialibrary/2011/04/07/81f83acc/CHPA0003%20Good%20practice%20guide%20to%20community%20heating%20and%20CHP.pdf.

CHPA. 2011b. *Leicester City Council*. London: Community Heat and Power Association (UK). Available at: www.chpa.co.uk/leicester-city-council_122.html.

DECC. 2011a. *Solar Thermal Water Heating*. London: Department of Energy and Climate Change. Available at: www.decc.gov.uk/en/content/cms/meeting_energy/microgen/solar_thermal/solar_thermal.aspx.

DECC. 2011b. *Air Source Heat Pumps*. London: Department of Energy and Climate Change. Available at: www.decc.gov.uk/en/content/cms/meeting_energy/microgen/ashps/ashps.aspx.

DECC. 2011c. *Great Britain's Housing Energy Factfile*. London: Department of Energy and Climate Change. Available at: www.decc.gov.uk/assets/decc/11/stats/climate-change/3224-great-britains-housing-energy-fact-file-2011.pdf.

DFPNI. 2010. *Case Study no. 12: CHP Installation at Ards Leisure Centre*. Belfast: Department of Finance and Personnel, Northern Ireland. Available at: www.dfpni.gov.uk/good_practice_case_study_no.12.pdf.

EST. 2005. *Potential for Micro-Generation: Study and Analysis*. London: Energy Saving Trust. Available at: www.energysavingtrust.org.uk/uploads/documents/aboutest/Micro-generation%20in%20the%20UK%20-%20final%20report%20REVISED_executive%20summary1.pdf.

EST. 2011a. *Ground Source Heat Pumps*. London: Energy Saving Trust. Available at: http://www.energysavingtrust.org.uk/scotland/Generate-your-own-energy/Ground-source-heat-pumps?gclid=CMW05JXTmqwCFYEZ4QodRV03Og.

EST. 2011b. *Hydroelectricity*. London: Energy Saving Trust. Available at: http://www.energysavingtrust.org.uk/Generate-your-own-energy/Hydroelectricity.

GEA. 2011. *Geothermal Basics*. Washington, DC: Geothermal Energy Association. Available at: www.geo-energy.org/geo_basics_plant_cost.aspx.

Kutscher, C. 2001. *Small Scale Geothermal Power Plant Field Verification Projects*. Golden, CO: National Renewable Energy Laboratory. Available at: www.nrel.gov/docs/fy01osti/30275.pdf.

MacKay, D.J.C. 2011. *Sustainable Energy – without the Hot Air*, updated online version. Available at: www.withouthotair.com. (Originally published in 2008 by UIT, Cambridge.)

Moss, J. 1982. Wave power teams braced for government's decision. *New Scientist*, 4 February. Available at: www.newscientist.com.

Mulliner, R. 2011. *Can Small Scale On Farm Anaerobic Digestion Be Commercially Viable?* Ludlow: Marches Biogas. Available at: www.rase.org.uk/events/agri-science-events/AD_Walford-Mulliner.pdf.

OIPAF. 2008. *The Oregon Solar Highway*. Salem, OR: Office of Innovative Partnerships and Alternative Funding. Available at: www.oregon.gov/ODOT/HWY/OIPP/inn_solarhighway.shtml.

REN. 2011. *Renewables 2011: Status Report*. Paris: Renewable Energy Policy Network for the 21st Century. Available at: www.ren21.net/Portals/97/documents/GSR/REN21_GSR2011.pdf.

Scottish Government. 2008. *Scottish Hydropower Resource Study*. Edinburgh: Scottish Government. Available at: www.scotland.gov.uk/Resource/Doc/917/0064958.pdf.

Suzuki, H., Dastur, A., Moffatt, S. and Yabuki, N. 2009. *Eco2 Cities: Ecological Cities as Economic Cities*. Washington, DC: International Bank for Reconstruction and Development/World Bank.

US DoE. 2006. *Geothermal FAQs*. Washington, DC: United States Department of Energy. Available at: http://www1.eere.energy.gov/geothermal/faqs.html.

US DoE. 2011a. *Heat Pump Systems*. Washington, DC: United States Department of Energy. Available at: www.energysavers.gov/your_home/space_heating_cooling/index.cfm/mytopic=12610.

US DoE. 2011b. *Geothermal Heat Pumps*. Washington, DC: United States Department of Energy. Available at: www.energysavers.gov/your_home/space_heating_cooling/index.cfm/mytopic=12610.

Wang, U. 2011. The rise of concentrating solar thermal power. *Renewable Energy World*, 6 June. Available at: www.renewableenergyworld.com/rea/news/article/2011/06/the-rise-of-concentrating-solar-thermal-power.

Weber, C. and Shah, N. 2011. Optimisation based design of a district energy system for an eco-town in the United Kingdom. *Energy*, **36**, pp. 1292–1308.

5 Carbon management in the new build

5.1 Defining the 'carbon problem'

The built environment sector remains a significant contributor of GHGs, and its contributions continue to grow in many countries. Ürge-Vorsatz *et al.* (2007) estimate 33 per cent of all global carbon emissions to be originating from existing buildings. In the UK, the residential sector alone contributed over 17 per cent of the total CO_2 emissions (see Figure 5.1). Part of the 'Public' and 'Business' sectors in Figure 5.1 consists of emissions from buildings. If these too are included, the total built environment consumption of energy is approximately 34 per cent of UK final energy (Ward, 2008). Based on these, UK energy consumption for space and water heating, cooking, lighting and appliances is responsible for 27 per cent of total carbon dioxide emissions (DECC, 2009; Ward, 2008). However, if we disaggregate the emissions by activities that are relevant to the built environment, nearly 45 per cent of all carbon emissions come from heating and moving air and water, and the use of appliances, in existing buildings (with a split between domestic and non-domestic buildings of 27 per cent and 18 per cent respectively; Kelly, 2009). The remaining 55 per cent is split between transport (33 per cent) and industrial processes (22 per cent).

Another key factor to note is that 87 per cent of all buildings that contribute to today's carbon emission in the UK will still be functioning in 2050 (Kelly, 2009) when the country is legally

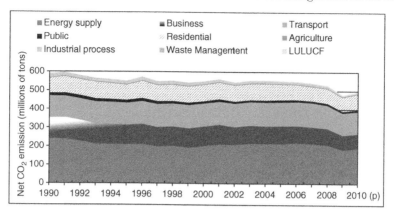

Figure 5.1 CO$_2$ emission in the UK in the recent past

Source: Drawn from data supplied by DECC, 2011.
Note: LULUCF = Land Use, Land Use Change and Forestry.

obliged to reduce its carbon emission level to less than 20 per cent of its emission in 1990. The absence of serious retrofit efforts in the existing stock will certainly contribute to the country missing its legal emission targets. The next chapter will examine the existing stock in greater detail.

Thus a low carbon built environment is a crucial linchpin of national carbon reduction attempts in any country. In order to be economically, socially and technically effective, our attempts at a low/zero carbon (LZC) built environment must be informed by the following 'principles for a low carbon built environment' (Grant *et al.*, 2009):

- Action focused on the most cost-effective carbon abatement measures should be a priority. Existing buildings present the main opportunities for cost-effective carbon abatement through energy conservation and efficiency.
- The prospect of early and sustained emissions reductions can justify the implementation of relatively costly carbon abatement options. New buildings offer a unique opportunity to implement relatively costly carbon saving measures that are more expensive or impractical to retrofit in existing buildings.

In other words, cutting edge but costly products, processes and technologies are likely to be most effective in new buildings, while existing buildings offer the most cost-effective quick wins. Both these approaches are necessary to ensure carbon savings are 'absolute' rather than merely 'relative' (as in efficiency improvements). At the same time, the increasingly stringent building regulations in many countries (including in the developing world – see Janda, 2009) are beginning to mandate LZC buildings in the case of new build. The existing stock on the other hand remains problematic. This is further confounded by increasing affluence, lifestyle changes oriented towards greater reliance on more, sophisticated and power hungry gadgets, and the so-called rebound effects. The current depressed state of the construction sector in several countries further adds to the pressure to do nothing to promote LZC buildings. Additionally the need to be equitable and just in our approach to LZC buildings too must be kept in mind.

5.1.1 Principal emission drivers in the built environment

Most of the carbon emission in the built environment (80–90 per cent) comes from energy use in buildings and cities for heating and cooling, electricity to power buildings, and transport (Table 5.1).

Table 5.1 Principal emission drivers in the built environment

Drivers	Fossil fuel combustion	Refrigerant leakage	Chemical processes	Land use changes
Building energy	▓			
Materials	▓			
Transport	▓		▓	
Waste	▓		▓	▓
Land				▓
Refrigerant		▓		

Source: Adapted from Grant *et al.*, 2009.

Of the total carbon in the built environment, 10–20 per cent is embodied in the building materials and fabric. The embodied carbon will be eventually released into the atmosphere at the end of the building lifecycle. Furthermore, the process of construction and the eventual demolition of the built environment directly drive land use changes which further add to the emissions from the built environment sector. Thus the key drivers from the built environment sector are:

- energy use;
- materials;
- transport associated with the built environment;
- waste;
- land use;
- refrigeration.

The six drivers listed above influence carbon emission to varying degrees at different stages of a building's lifecycle. Technologies and processes for carbon reduction in buildings will also depend on the pathways through which they attempt to reduce energy use: supply improvement (i.e. low carbon energy supply), demand reduction, efficiency improvement or better energy management (see Table 5.2).

5.2 Physics of buildings

In order to understand the GHG emissions from the built environment, it is essential to understand the energy and material flow through buildings, as well as processes such as comfort, lifestyles and function.

5.2.1 Climate

The climatic context of the built environment is centrally important to its energy and thermal performance. For the purpose of energy use in buildings, the Köppen–Geiger classification system (Köppen, 1923) as modified by recent science presents the most comprehensive classification of the world's climate (see Figure 5.2). The Köppen–Geiger classification system divides the world into five broad climate types, further subdivided according to precipitation and temperature (limiting factors):

- A – tropical, rainy climate;
- B – dry climate;
- C – humid, mesothermal climate;
- D – humid, microthermal climate;
- E – polar climate.

Table 5.2 Principles of carbon management in the built environment

Lifecycle stages	Options for LZC buildings			
	Supply side options	Demand reduction	Efficiency improvement	Energy management
Site selection	On-site micro-generation of renewable energy			
Site layout and planning	Integration of micro renewables	Effective use of sunlight*		
		Create good microclimate*		
		Efficient built form*		
Design, construction and refurbishment		Insulated/air tight envelope*	Efficient fixed service equipment	Effective metering
		Avoidance of mechanical systems*	Effective controls	Operations and maintenance instructions
Occupation, operation and management		Effective use of controls	Adequate maintenance of equipment	Feedback on energy use
		Adequate maintenance of envelope and fabric		

Source: Modified from Grant *et al.*, 2009.

Note: * These strategies are often grouped together as 'passive design techniques'.

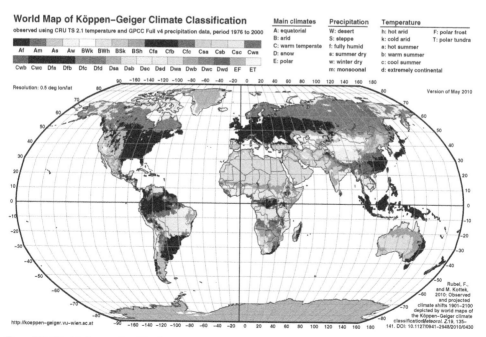

Figure 5.2 World climate classification

Source: Rubel and Kottek, 2010.

Table 5.3 Köppen–Geiger classification flow and areas covered by each climate type

Climate type/ subtype	Description	Climate criterion	Continental area of the world covered by type/subtype (%)
A	Tropical rainy climate	$t_{min} \geq 18°C$	22.6
Af	Tropical rainforest	$r_{min} \geq 6cm\ mo^{-1}$	3.2
Aw	Tropical savannah	$r_{min} < 6cm\ mo^{-1}$	19.4
B	Dry climate	$\bar{r} \leq r_d$	15.1
BS	Steppe	$\bar{r} \geq r_d/2$	5.7
BW	Desert	$\bar{r} < r_d/2$	9.4
C	Humid meso-thermal climate	$-3°C \leq t_{min} < 18°C$	19.1
Cs	Warm climate with dry summer	$r_{wmax} \geq 3\ r_{smin}$	2.3
Cw	Warm climate with dry winter	$r_{smax} \geq 10\ r_{wmin}$	11.1
Cf	Humid temperate	$r_{smax} < 10\ r_{wmin}$ and $r_{wmax} < 3\ r_{smin}$	5.7
D	Humid micro-thermal climate	$t_{min} < -3°C$ and $t_{max} \geq 10°C$	23.7
Dw	Cold climate with dry winter	$r_{smax} \geq 10\ r_{wmin}$	19.4
Df	Cold climate with moist winter	$r_{smax} < 10\ r_{wmin}$	4.3
E	Polar climate	$t_{max} < 10°C$	19.6
ET	Tundra climate	$0°C \leq t_{max} < 10°C$	9.6
EF	Permafrost	$t_{max} < 10°C$	10.0

Source: Lohmann *et al.*, 1993.

Table 5.3 presents the limiting factors (precipitation and temperature) that distinguish the climate zones and the fraction of the continental area of the world covered by each type.

The weather variables that are of importance to building physics (and therefore its energy and carbon consequences) are:

- *Air temperature:*
  - dry bulb temperature (°C);
  - wet bulb temperature (°C).

- *Air humidity:*
  - humidity ratio – either vapour pressure (kPa) or absolute humidity (g m^{-3});
  - relative humidity (%).

- *Wind:*
  - speed (m s^{-1});
  - direction (°).

- *Radiation:*
  - direct/diffused solar radiation (W m^{-2});
  - longwave.

The effects of these variables on buildings are further influenced by geographical features such as latitude, elevation and distance from the sea. Latitude determines the type of heat load (heating or cooling and/or both), while elevation above the sea leads to local cooling (at an approximate rate of 1°C per 100 metres elevation). Distance to the sea determines the magnitude of annual variations in local climate caused by the landmass (the so-called 'continentality' effect).

The effects of climate variables on building energy needs can be determined either by simple rules of thumb or by more exhaustive computer simulations. Data needs for each of these types of methods will vary:

- *Simple methods:*
  - monthly averages;
  - seasonal (winter/summer) averages.
- *Detailed methods:*
  - typical reference year (TRY);
  - hourly values;
  - actual measurements.

5.2.2 Indices to quantify the climatic burden on buildings

Efforts to quantify normative energy consumption in buildings (especially housing) began with the first energy crisis in the 1970s. Given the wide disparity in energy consumption patterns even within the developed world and the importance of the built environment in reducing national energy consumption, there was a need to benchmark building energy performance with a view to standardising and codifying the best practices. A review of such early attempts is given by Yannas (1994).

An early 'energy index' for temperate climates was developed by Yannas (1990). According to this index (the Energy Index – EI), a notional detached dwelling complying with the 1990 UK building regulations would consume approximately 100–115 kWh per annum per square metre (m^2) of floor area. A building fulfilling all the then known 'good design' principles (i.e. south facing windows, double glazing, insulated walls and roofs, tight construction and mechanical ventilation with heat recovery) will have an EI of less than 30 kWh/year/m^2 (Yannas, 1996). This compares well with the most stringent current standards, for example the Passivhaus standards (Feist, 2008) or the Code for Sustainable Homes (DCLG, 2006).

Attempts to quantify the 'climate burden' imposed by external climate on buildings have an even longer history. Early attempts from the time of Silpasastra in India (Acharya, 1979) and Vitruvius (Morgan, 1960) in Rome focused on design exemplars based on climate types. An approach based on meteorological data was attempted by Mahoney in 1965 (see Koenigsberger *et al.*, 1974).

A more robust yet simple approach was recently developed by Integrated Environmental Solutions Ltd (IES), the developers of building energy simulation software Virtual Environment (VE) (IES, 2011). The index, called the Climate Energy Index (CEI), attempts to quantify the 'climatic burden' on a building by outside air. It has four component loads: two sensible energy loads (heating and cooling) and two latent energy needs (humidification and dehumidification). The CEI is defined thus:

CEI = Sum of (Sensible Cooling, Sensible Heating, Humidification, Dehumidification)

It is expressed in kWh/yr for a given volume of air (in m^3/hr). Table 5.4 presents the CEI values for 14 geographical locations from around the world representing different climate zones.

Table 5.4 CEI values for selected representative locations

Location	CEI	Sensible heating	Sensible cooling	Dehumidification	Humidification
Fairbanks	31.96	26.83	0.00	0.00	5.12
Minneapolis	19.25	16.01	0.39	0.35	2.49
Boston	14.19	12.18	0.14	0.06	1.81
Baltimore	14.11	11.08	0.79	0.79	1.44
Glasgow	13.40	13.04	0.00	0.00	0.36
London	10.11	9.97	0.00	0.00	0.13
Los Angeles	5.06	4.85	0.00	0.00	0.21
Sydney	4.72	4.28	0.27	0.14	0.03
Phoenix	6.26	3.18	2.44	0.04	0.60
Houston	12.35	6.54	2.75	2.88	0.19
Abu Dhabi	14.77	3.35	5.94	5.48	0.00
Miami	12.01	4.05	4.13	3.83	0.00
Bangkok	22.43	4.79	8.05	9.58	0.00
Singapore	25.72	5.43	7.56	12.73	0.00

By normalising the CEI for building parameters, it is then possible to derive a Building Energy Index as below:

$$BEI = CEI \times \frac{OA}{FA}$$

where

BEI =   Building Energy Index (kWh yr^{-1} m^{-2})

OA =   Outside air intake, both infiltration and auxiliary ventilation (m^3 hr^{-1})

FA =   Floor area (m^2)

Dry climates with moderate or little precipitation and mild winters (such as Phoenix and Los Angeles, USA, and Sydney, Australia) are among those with the lowest 'climate burden', whereas equatorial monsoon or dry climates (such as Singapore and Bangkok, Thailand) are among those with the highest climate burden anywhere in the world outside the extreme polar environments.

The reason for such variations in 'climate burden' is more readily apparent when the weather parameters are plotted on a psychrometric chart. The psychrometric chart combines six climate parameters to enable the study of the moisture properties of air: dry bulb temperature, wet bulb temperature, absolute humidity, relative humidity, enthalpy and specific volume. Of these, the critical parameter in terms of building energy use is enthalpy (the internal energy in a parcel of air). Climates requiring no changes in enthalpy to move them into the 'comfort' zone will require the least amount of energy expenditure. The rest of the regions in the psychrometric chart require one or more of the following design approaches to make building interiors comfortable:

• *Passive design approaches:*

  ○ evaporative cooling;
  ○ inertia of the building envelope;
  ○ internal heat gains from human activities;
  ○ ventilation;
  ○ thermal inertia combined with night ventilation.

- *Active design approaches:*
  - dehumidification;
  - heating;
  - air conditioning.

Figure 5.3 show the monthly climate data for four cities from around the world plotted on a psychrometric chart. Figure 5.4 shows the BEI (energy loads) based on a standard building in 14 different locations around the world.

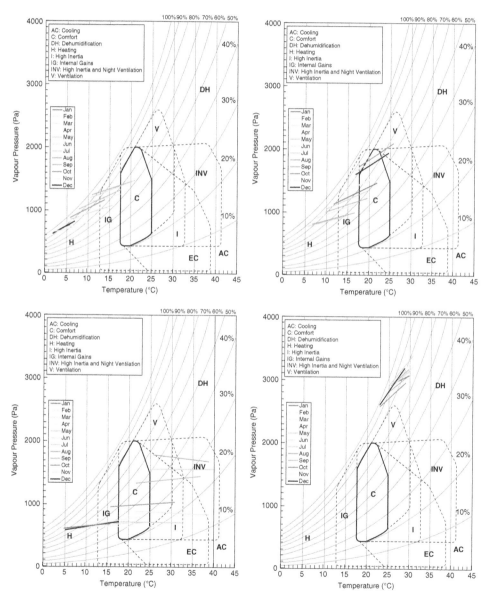

Figure 5.3 Climate and thermal comfort conditions for four representative cities around the world: London, Sydney, Phoenix, Singapore

Source: Based on climate data from www.weatherbase.com, plotted on an Excel workbook developed by H. Rosenland.

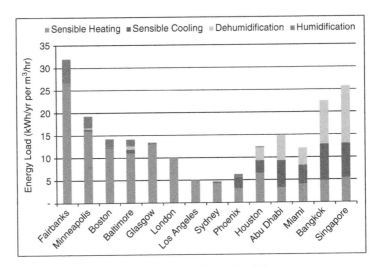

Figure 5.4 Variations in energy loads on buildings according to local climate

Source: Emmanuel *et al.*, 2012.

5.2.3 Thermal comfort

The climate parameters responsible for driving human thermal comfort are:

- air temperature;
- relative humidity;
- air movement;
- solar radiation.

Together with the above four climate parameters, clothing worn by humans and their activity levels determine the thermal comfort. These are combined in numerous ways to affect human comfort.

Human deep body temperature (the so-called 'core temperature') must be maintained at about 37°C for health. Since metabolised food releases energy, a healthy human body normally attempts to lose heat to the ambient environment at all times. The body can – for a short duration – gain heat, although this is not desirable for longer periods of time.

Under moderate environmental conditions, a human body immersed in the environment for at least an hour strives to achieve thermal balance. This steady-state heat flow per unit area per unit time is given by Fanger (1970: 22–23):

$$H - E_d - E_{sw} - E_{re} - L = K = R + C$$

where
H = internal heat production in the human body
E_d = heat loss by water vapour diffusion through the skin
E_{sw} = heat loss by sweat evaporation
E_{re} = sensible heat loss by evaporation
L = latent heat loss by respiration
K = conduction from outer surface of clothed body

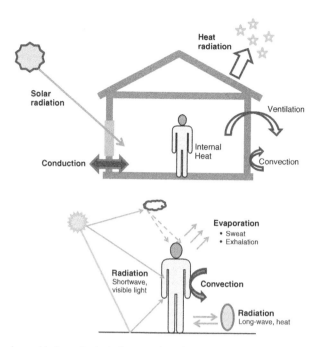

Figure 5.5 Human thermal balance in the indoors and outdoors

R = radiation loss from outer surface of clothed body
C = convection heat loss from outer surface.

H is a function of the level of human activity. The higher the activity the more the heat loss needed for thermal balance. E_d is a function of the difference between saturated vapour pressure at skin temperature and the actual vapour pressure in ambient air (Fanger, 1970). In warm and humid climates heat loss due to E_d is negligible.

L is a function of the humidity ratio difference between the inhaled and the exhaled air. L will be low in highly humid regions. E_{re} is usually low for all environments (Fanger, 1970: 29). Unless the body is in a reclined position, conduction loss (K) too is usually negligible.

At 'comfortable' air temperatures (20–25°C) and with little wind movement (< 0.1 m s^{-1}), the metabolic heat production is mainly dissipated by radiation (R), accounting for up to 60 per cent of heat loss. Convection loss at this stage will account for 15 per cent of the heat produced, while evaporation from the lungs and through the skin (E_{sw}) accounts for about 25 per cent heat loss (Oke, 1987). However, an increase in wind flow will proportionately increase evaporative and sensible heat losses and could lead to greater comfort. The increase in 'comfortable' temperatures due to wind is given by the formula $6V - V^2$, where V is the wind velocity (m s^{-1}) (Szokolay, 1992). For example, a wind speed of 0.5 m s^{-1} will increase the 'comfort' temperature by 2.75°C.

In warmer environments (air temperature >26.5°C), the human thermoregulatory mechanism cannot depend on radiation, for radiation is a function of the temperature *difference* between the source and the sink (in this case, between a human body and its surrounding environment). Thermoregulation by convection (*C*) too becomes insignificant if ambient air temperature rises above 35°C. At this point, the air is warm enough to act as a heat *source* to the body rather than a sink (Oke, 1987: 223). Therefore the body dilutes the blood vessels just under the skin in an attempt to reduce the insulation potential of the skin (thus the skin becomes red with exposure to a warm environment) (Fanger, 1970). By diluting the blood vessels, more warm blood flows very near the surface of the skin, and is able to lose some of its heat to the outside environment via radiation and/or convection. However, when the air temperature crosses the 35°C threshold, even this mode of heat loss becomes insignificant. Therefore, at higher ambient air temperatures, heat loss by humans must depend on evaporation (E_{sw}).

Monteith (1973) stated that a healthy adult male can lose up to 1 kilogram of water every hour. This loss translates into an hourly energy loss of 375 W/m². Although such large quantities of heat loss should be sufficient to cool the body in warmer environments, the actual evaporation in human beings is somewhat less owing to the increasing presence of salt on skin deposited by the evaporating sweat. Prolonged evaporative heat losses of higher rates are lethal. When an average man loses 2 per cent of his body weight equivalent of water (about 0.72 kilograms or 1.6 lb) by evaporation, he becomes thirsty. At 4 per cent loss he feels apathetic and impatient, at 8 per cent speech becomes difficult and, when the water loss reaches 18–20 per cent of body weight equivalent, death is imminent (Oke, 1987: 225).

Culturally, therefore, humans have attempted to enhance the thermoregulatory processes through built envelope manipulations, clothing modifications and maintaining a rhythmic balance of activities in conjunction with the climatic seasons. Lightly clad, open form buildings are the traditional norm in warm climates, while tightly closed, heavy constructions are common in cooler climates. In terms of clothing, the warm climate custom is to utilise loosely woven, thin materials covering a bare minimum of the body, while in cooler climates layered clothing of high insulation value worn with heavy underclothes is common (Mather, 1974). Similarly, in terms of activities, traditional societies in the tropics have developed working habits beginning earlier in the day when the sun is low, or late in the evening, soon after sunset (Correa, 1989). These actions, which we will later call adaptive behaviour, enable people to tolerate a wide range of ambient conditions through the manipulation of the cultural environment as well as the thermoregulatory processes.

5.2.4 Key building processes and needs

The key to LZC buildings is an understanding of the processes that drive human comfort in the built environment and the climatic context in which a building is situated.

The building functions that could be manipulated to derive comfort are: air movement, building thermal performance and moisture. These must be balanced against the need to account for aural and visual comfort and accommodate bodily biological processes (RAE, 2010).

Air movement

Adequate fresh air supply is essential for the occupants of buildings, but air movement carries with it humidity, heat (or cold, depending on outdoor climate conditions), pollutants and sound. Air movement is driven by pressure and temperature differences through flow paths. Controlling the air movement so as to achieve warmth (reduced air flow to minimise heat loss) or coolth (increased air flow to minimise heat gain) is needed depending on the outside climatic context. Proper understanding of air movement is also necessary to manage mechanical ventilation systems efficiently where needed.

Thermal performance

The building envelope performs the vital function of modulating the thermal vagaries of external climate (by augmenting warmth in cooler climates and preventing heat gain in warmer ones). The variables that matter the most are:

- thermal transmission (U-value) through the building envelope – always needs to be lower, to prevent heat gain in warmer climates and heat loss in cooler climates;
- building openings – smaller in cooler climates to reduce heat loss but larger in warm climates to encourage heat loss;
- thermal mass/insulation – high/heavy to delay heat loss in cooler climates but low/light to encourage a quicker response to outdoor climate in warm regions.

Control of moisture

Moisture is introduced into buildings from the environment, from the breath of its occupants and from the transpiration of plants and animals. Excess moisture can result in problems of condensation, leading to the growth of mould and the development and persistence of odours. Moisture is also the primary agent of deterioration in buildings, and hence its control is essential to ensuring the durability of structures. Moisture moves by a number of mechanisms: capillary flow, vapour diffusion, air convection and gravity flow. Measures for controlling the build-up and transport of moisture within both the interior and the fabric are essential for the health and well-being of building occupants.

Acoustics

The basic physics of sound propagation is simple, but the interaction of sound pressure waves with complex shapes and multi-layer constructions with openings, as in buildings, is more challenging. Controlling noise, both from the internal and external environment and from the internal mechanical and electrical services in buildings, is essential to create environments that promote aural communication and comfortable working conditions.

Light

Light is essential for function, but simply providing sufficient illumination by electric lighting is rarely adequate for high performance buildings. Lighting design must consider source intensities, distribution, glare, colour rendering and surface modelling if we are to create stimulating, high quality interior environments. Daylight is often dismissed in lighting design as being too variable to be reliable, but daylight design is essential to reduce reliance on artificial lighting.

Biology

In addition to the fundamental physical aspects of building design, LZC buildings must also respond to human physiology, particularly relating to comfort and task performance. A basic understanding of biology and ecology creates opportunities to enhance the natural environment and supplement the performance of the building through the integration of planting and landscaping. Planted roofs and shading by deciduous trees both make valuable contributions to the thermal performance of buildings (RAE, 2010).

5.3 Passive/low energy design approaches

Heating and cooling loads can be reduced through ventilation, heat sinks, the use of solar and other natural heat sources, and improved insulation, windows and equipment. Power loads can be reduced through improved lighting (e.g. LED, compact fluorescent light bulbs and increased use of natural lighting) and the use of energy efficient appliances. Integrated building design and the modification of building shapes, orientation and related attributes can also reduce energy demand, as can changes in energy-wasting behaviour and improved operations and maintenance (Levine *et al.*, 2007). The technologies for these efficiency improvement measures are commercially available and have been validated through their use in contemporary buildings. The energy saving potential in the building sector is large. The Fourth Assessment Report of the IPCC, based on the results of over 80 surveys worldwide, concluded that there is a global potential to reduce approximately 29 per cent of the projected baseline emissions from residential and commercial buildings by 2020 and 31 per cent from the projected baseline by 2030 at a net negative cost. The potential is the highest and cheapest among all sectors studied (Levine *et al.*, 2007). The exact approaches to lowering carbon in buildings depend on the climatic context in which the buildings are located.

5.3.1 Temperate climates

The LZC design process appropriate for temperate climates can be summarised into three steps (Wang, Gwilliam and Jones, 2009): analyse local climate data to make use of the local climate condition for promoting LZC buildings; apply passive design methods and advanced façade designs to minimise the load requirement from heating and cooling through building energy simulations; and use renewable energy systems including photovoltaic, wind turbines and solar hot water systems to convert LZC building to a net exporter of energy.

Noting that building energy demand and the subsequent emissions of CO_2 are the expression of a complex and highly interdependent web of many socio-technological networks from government to utilities, to builders, to the individual consumer (see Chappells and Shove, 2003), Monahan and Powell (2011) suggest the following as the framework towards LZC in temperate climates:

- Reduce the need for energy inputs (e.g. increase levels of insulation, reduce unwanted ventilation, and design strategies that optimise solar gain – termed passive solar).
- Decarbonise grid electricity fuel systems and change the way energy-dependent services are provided. Displace fossil fuels with alternative, renewable energy sources and new low or zero carbon technologies (e.g. solar hot water, photovoltaics, wind, hydro and biomass).
- Increase the efficiency of service provision (e.g. A* rated boilers or heat pumps).
- Influence consumer behaviour to induce change in the desired, low energy direction (e.g. provide information such as product energy labelling and government funded social 'marketing' campaigns).

One of the approaches that combines many of these strategies applicable to temperate climates is the so-called 'Passivhaus' approach. Although originally developed for dwellings, this strategic approach can be used for any type of buildings in temperate climates. Its applicability in warm climates (especially warm, humid climates) is limited.

Case study: The Passivhaus approach

The Passivhaus concept was first developed by Bo Adamson and Wolfgang Feist in 1988. The first Passivhaus buildings were built in Darmstadt in 1990, and the Passivhaus Institute was founded in 1996. Since then more than 15,000 Passivhaus buildings have been built worldwide, most of them in Germany and Austria. There are now a number of other Passivhaus standards, such as Minergie-P in Switzerland.

A Passivhaus is 'a building in which thermal comfort is guaranteed solely by re-heating (or re-cooling) the fresh air that is required for satisfactory air quality' (NBT, 2009). The principle of the Passivhaus is a building with no heating system except for the heat recovery through the ventilation unit. If built correctly, the building is robust, healthy and cost-effective. To achieve a Passivhaus, strict design guidelines must be followed in relation to:

- compact design;
- high levels of insulation;
- minimal thermal bridging;
- highly insulating windows;
- very high standards of air tightness;
- high quality mechanical ventilation with heat recovery (MVHR).

A building envelope should be air tight when all ventilation openings are closed. The design requirement for air changes has to be provided by opening the windows manually, other controllable ventilation openings or suitable mechanical ventilation systems.

When assessing the air permeability of the building envelope, the following aspects must be considered separately:

- Individual building components must exhibit the necessary air tightness in accordance with building component standards.
- The overall air permeability of the building envelope must meet the limiting and target values of building regulations.
- Local air permeability (leaks, primarily on the inside) can lead to moisture damage, because they allow moist interior air to infiltrate the construction.
- Local air permeability and associated draughts can have a detrimental effect on the thermal comfort of the occupants and can also lead to increased energy consumption.

General principles

Design considerations

- Compact building form: the ratio between the building surface and the building volume should be as low as possible (dormers, bays and so on are better avoided).
- Building orientation: typically the building is oriented toward the south, where the façade of a Passivhaus has many big windows. The north façade has only a few small windows. This leads to big solar gains.
- To avoid overheating in summer the south façade should contain a sun screen.

Thermal bridges

U-values below 0.15 W/m^2K require the prevention of thermal bridges. As the U-value decreases, thermal bridges become more and more significant:

- The proportion of timber over the whole cross-section should be minimised (timber conducts heat approximately four times more than insulation materials).

- Windows should be embedded into the insulation layer.
- The detailing at all junctions and joints (corners, windows, doors, plinths, suspended floors and so on) requires special attention.
- Simple architecture is favourable.
- Built-in roller shutters should be avoided.

Moisture control

Building practice dictates that the diffusion resistance of the individual layers must decrease from the inside to the outside.

An effective vapour control layer on the inside of the structure (e.g. wood based products such as OSB) reduces the diffusion of water vapour. In combination with the vapour open PAVATEX boards, this ensures that no damaging amounts of condensation build up within the construction (interstitial condensation):

- The vapour control layer must lie on the warm side of the thermal insulation.
- The vapour control layer must cover the surface of the entire building envelope.

The vapour control layer can be combined with other component layers, generally with the air tightness layer.

Air tightness

Thermally insulated constructions require a permanent air tight layer on the inside:

- To avoid damage to the construction and to prevent heat loss the air tightness layer must be installed very carefully, especially at junctions between components, at joints between elements, around penetrations etc.
- Fewer penetrations will allow for simple, cost-effective construction.
- Pipes and cables should not damage the air tightness layer in any way.

The air tightness layer is more important than the vapour control layer in terms of preventing damage to the fabric of the building.

Ventilation cavities

The primary function of ventilation cavities behind the cladding and/or the roofing material is to allow the air flow to carry away any moisture present by way of convection. Whether the moisture is a result of vapour diffusion, precipitation or wet building trades is irrelevant. Ventilated constructions (external walls with cladding, ventilated roofs) are regarded as favourable from a diffusion viewpoint, and do not require the diffusion behaviour to be verified by calculation, provided the moisture loads do not exceed those equivalent to normal residential and working situations.

Services

The production of hot water is the component with the highest energy consumption in a Passivhaus:

- Pipe runs should be as short as possible and well insulated. Dead legs should be avoided.
- Pipes and cables must lie on the warm side (inside) of the insulation.
- Low water appliances should be considered.

Ventilation and heating

Passivhaus solutions can be considerably more air tight than conventional constructions, achieving air change rates smaller than 0.6 m^3 per m^2 at 50 Pa. Healthy buildings require a minimum air exchange rate of 0.5 air changes per hour at 50 Pa. Therefore additional ventilation is a sensible and beneficial addition to a building constructed according to the Passivhaus standard. MVHR systems bring controlled volumes of fresh air into all rooms of the building and remove a controlled volume of moisture laden or stale air to the outside. With the heat recovery system these units can recover heat from outgoing air to preheat the incoming air. This heat recovery can provide a large proportion of the heat required to keep the building at comfortable living temperatures. However, MVHR with integrated economic auxiliary heating is also available.

It is essential that the MVHR system is specifically sized and that the environmental and improved indoor air quality advantages are associated with all systems.

- Under normal circumstances a standard sized house is supplied by the MVHR with about 100 m^3/h of fresh air to the living and sleeping rooms.
- In special needs it can be set to a higher setting where between 160 and 185 m^3/h is provided.
- The same quantity of charged or polluted air is being sucked away in wet areas such as the kitchen, bathroom and shower.

Figure 5.6 An example of the Passivhaus approach
Photo credit C. Morgan.

Case study: Zero carbon domestic buildings in temperate climates

Monahan and Powell (2011) compared four approaches to low/zero carbon buildings in the temperate climate region of south-eastern England (Lingwood, Norfolk). The four cases were:

1   a control house fulfilling current building regulations, with instantaneous gas fired heating and hot water, with no additional renewable technology (CONTROL) (see Table 5.5 for building envelope details, air tightness and assumed energy demand);
2   as above, plus grid connected PV and solar hot water systems supplementing hot water and electricity (SOLAR);
3   as above, plus passive solar sunspace and MVHR;
4   as above, plus all electric space heating and hot water provided by a 3.75 kW ground source heat pump with an under-floor heating loop installed on the ground floor only with radiators upstairs (ground source heat pump).

Table 5.5 Thermal parameters of the 'CONTROL' home fulfilling current UK building regulations

Parameter	Value
U-value wall	0.18 W/m^2 K
U-value floor	0.16 W/m^2 K
U-value roof	0.14 W/m^2 K
U-value window	1.80 W/m^2 K
U-value door	2.40 W/m^2 K
Air permeability	7.00 m^3/m^2 h at 50 Pa
Heat loss	1.33 W/m^2 K
Emission	22.30 kg CO_2/m^2 year
Space and water heat demand	50.00 kWh/m^2 year

Results in Figure 5.7 show that all technological options (ground source heat pump, MVHR and solar PV) are useful in reducing emission and energy use in buildings. They are also more or less cheaper to run than a conventional home. Solar PV gave the most improvement in terms of both emission and energy use reduction as well as running cost. The heat recovery approach as advocated by the Passivhaus strategy did not lead to substantial savings in running costs (approximately 14 per cent lower than the CONTROL house – Monahan and Powell, 2011). Heat pumps were found to have a relatively poor performance in terms of primary energy demand. They were also found to have comparatively high carbon emissions and running costs. What this points to is the urgent need to 'decarbonise' the electricity grid even as the building envelope solutions aim to reduce carbon in buildings. What is clear from this study is that the reductions in both energy and consequential carbon emissions were derived principally from the increased thermal efficiency of the homes and consequent reduction in heating-related energy demand (Monahan and Powell, 2011). Another point is the relatively low differences in non-heating energy demand, pointing to the importance of user behaviour.

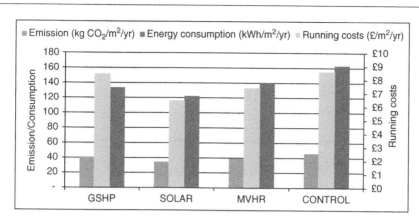

Figure 5.7 Average emissions, primary energy use and running costs of the case study houses

Source: Based on data from Monahan and Powell, 2011.

Case study: The case of zero energy (i.e. energy exporting) buildings in the UK

The case of an energy exporting house using solar PV panels and wind turbines was examined by Wang *et al.* (2009). This study simulated the energy consumption and renewable energy generation for a two-storey single-family house under Cardiff, UK weather conditions. Table 5.6 shows the envelope parameters, design load and annual energy consumption, as well as annual renewable energy generation.

Table 5.6 Parameters of a zero energy building for the UK

Parameter	Value
Building envelope:	
Window: wall ratio	0.4 (south façade)
	<0.1 (all other façades)
U-value	External wall = 0.1 W/m² C
	Roof = 0.2 W/m² C
	Glazing = 1.367 W/m² C
Space heating:	
Under-floor heating combined with	Active layer pipe grid, pipe spacing 100 mm centre to centre; dia = 20mm.
heat pump system	Pipe wall thickness = 2 mm.
	Pipe wall conductivity = 1.26 kJ/h m K.
Domestic hot water	Load = 98 L/day.
	Solar collector area = 5 m² mass flow rate = 20 kg/h.
	Solar collector efficiency = 35%.
Renewable energy systems:	
Solar PV	Rated power = 165 W.
	Area = 1.26 m².
	Array = 4 × 2.
	Total rated power = 1.32 kW.

Wind	Two numbers 2.5kW wind turbines with a hub height = 15m.
Annual electricity consumption:	
Lighting and appliance	−4,672.0
Auxiliary heating in SDHW	−401.7
Space heating	−935.2
Annual PV generation	+687.8
Annual wind turbine generation	+6,618.1
Net generation	+ 1,297.0

Source: Based on data from Wang *et al.*, 2009.

Note: * These strategies are often grouped together as 'passive design techniques'.

While it is apparent that the two renewable energy generation systems used (solar PV and wind) are capable of meeting the annual energy demand, it must be noted that the energy generation profile does not match the demand profile. A theoretical possibility exists in the UK to make a zero energy house similar to the one reported here. However, in practice a system of net metering (import–export electricity meter) to manage the two-way energy flow from the building and a grid capable of synchronising a multitude of mini-energy generation sources without significant power loss in transmission is needed to make such buildings truly zero energy. Additionally, the overwhelming reliance on wind power (nearly 91 per cent of all on-site generation) is difficult to maintain in urban areas.

5.3.2 Warm, humid climates

The primary goal of an LZC building in warm climates is to restrict solar gain, i.e. it is shading. Every other design decision must facilitate shading and augment the cooling so obtained by other means. Ventilation could play an important part in this regard. Night-time comfort must receive special attention in the tropics. Since nights in the tropics are typically too warm for comfort, (Nieuwolt, 1986: 532), heat storage in the built mass must be kept minimal.

Givoni (1989) suggested the following as key design approaches to building design in warm, humid climates, both at the individual building scale and at the neighbourhood level:

1   street layout and system of network;
2   density of built-up area;
3   average and relative height of buildings;
4   proper choice of building types (thick, thin, doughnut, etc.);
5   design details of green spaces (Givoni, 1989: 7–6).

Given the light thermal mass and relatively open nature of buildings in warm, humid climates, much of the LZC approach needs to originate at a scale larger than individual buildings so that shading and ventilation can be adequately exploited. It is therefore more important to manipulate street geometry than other design factors like surface material type, colour and so on, at least in the small to medium scale. While wind movement is crucial to thermal comfort in warm, humid climates, it must be remembered that warm, humid wind movement is low and unpredictable (typically lower than 1 m/s; Nieuwolt, 1986). At such low speeds, even small obstacles like trees could significantly alter wind flow. Furthermore, the cooling effect of wind movement, which is proportional to the difference between skin temperature and air temperature, is of little value in the tropics, where both these temperatures are almost equal: 35°C. To complicate matters even more, the tropical wind patterns keep shifting in the onslaught of a monsoon.

Bio-climatic influences of vegetation in warm climates

The effect of trees on individual buildings is impressive. Air temperature reductions of up to 11°C have been recorded (Miller, 1988: 53). Home cooling costs are known to have gone down by as much as 50 per cent in Florida (Parker, 1983). Part of the reason why vegetation is so effective in cooling is the porosity of the tree canopy (Heisler, 1974). A large amount of radiation impinging upon vegetation is retained to be used for transpiration. About 10–25 per cent is used to heat the air. In contrast, open, paved areas like car parks absorb up to 85 per cent of the incident solar radiation, and the rest is directly used to heat the air. This absorbed heat too will eventually heat the air. Myrup (1969) has estimated that areas with no vegetation would partition 40 per cent of net all-wave radiation into sensible heat and the balance into heat storage, which in turn will add to sensible heat gain after sunset. Areas with a significant tree population, on the other hand, partition only 10 per cent of net all-wave radiation into sensible heat. Another 10 per cent is stored in the biomass, while the rest is partitioned into latent heat loss. Evaporation by vegetation is much higher, since the leaves of a tree crown usually have five to seven times more area than the maximum horizontal cross-section of the tree crown (Leonard, 1972).

Using the stored energy, vegetation can also transpire efficiently. Even an isolated tree can transpire up to 380 litres of water per day (Kramer and Kozlowski, 1960). This is equivalent to the cooling provided by five 20,000 BTU (5,800 W/m²) air conditioners running 20 hours per day (Federer, 1971). What trees provide therefore is relief from solar radiation by a combination of absorption and evaporation, not so much relief from air temperature per se. Although trees transpire a significant amount of water, this water does not appear to be destructive to human comfort by increasing relative humidity (Federer, 1971).

Another design element of use in a warm, humid climate is water. Though humidity can be increased by the presence of large water bodies, decreased air temperatures in built-up areas more than compensate for this. The cooling potential of water, if properly shaded and located on a major wind path, is immense, even in the tropics (Emmanuel, 2005).

Table 5.7 Effect of vegetative cover on heat partitioning

	Rural	*Suburban*	*Urban*	*Urban centre*
Green:built ratio	100:0	50:50	15:85	0:100
Energy (W/m²)				
$Q^* + Q_F$	535	554	546	530
Q_{HO}	150	216	240	370
Q_E	305	216	158	0
ΔQ_S	80	122	148	160
Heating rate (k/w)				
Sensible heating	0.5	0.8	0.9	1.3
Evaporative suppression	0.9	0.7	0.5	0
Net change	−0.8	−0.6	−0.5	−

Source: Oke, 1989: 343.

Note: Q^* = net radiation; Q_F = anthropogenic heat; ΔQ_S = surface heat storage; Q_{HO} = surface turbulent sensible heat; evaporative suppression = thermal equivalent of energy used in evaporation which would otherwise contribute to turbulent warmth; net change = difference from bare city case.

Principles of bio-climatic buildings in warm climates

RESTRICT SOLAR RADIATION

Solar shading depends on the following:

1   location;
2   orientation;
3   time/day;
4   building/site dimensions.

Shading depends on the time(s) a designer wants to avoid sun. In the case of the tropics, the sun's heating pattern offers a starting point in determining the time the sun needs to be avoided, in other words the extreme limits of time the sun's rays must be avoided: the cut-off time. The cut-off time will have a starting time (the time *until* which sun may be let in) and a finishing time (the time *after* which sun may be let in).

Based on these principles Emmanuel (2005) presented a method to work out the shading envelope (called the 'shadow umbrella') needed to restrict solar radiation in warm places. The principles of shadow angles lead to the following general building and neighbourhood form determinants for warm, humid climates:

1   The most preferred orientations are 0°–80° (i.e. within the north-eastern quadrant).
2   The urban density should be one that slopes towards the north-east (i.e. smaller buildings in the north-east and taller ones towards the west).
3   The north-eastern and south-western corners of a development could be left open.

PROMOTE VENTILATION

The task of designing for wind movement should begin at settlement level, particularly in its street layout pattern. Streets should be so aligned as to reap the maximum benefit from macro-level wind directions. In the tropics, this would usually mean aligning the streets along the major monsoonal wind directions. Additionally, the following considerations will help:

1   location of a town within a region;
2   density of the urban area;
3   orientation and width of streets;
4   heights and relative heights of buildings (Givoni, 1994: 1048).

Building densities and heights are another important variable in promoting ventilation in the tropics. Usually high density zones are prone to poor ventilation regimes, and long walls of building fabric prevent deeper penetration wind. Givoni (1994: 1050) suggests that 'an urban profile of variable building height, where buildings of different heights are placed next to each other, and when the long facades of the buildings are oblique to the wind enhances urban ventilation'.

In the weak wind regime of the warm, humid tropics, another possibility is to induce wind flow by the thermal differences that arise at the edges of water bodies. Differences in the thermal properties of land and water generate sea/land breezes at day/night respectively. These wind flow patterns could be effectively used by sensitive urban planning measures that promote deep wind penetration into cities.

PROMOTE EVAPORATIVE COOLING

While the direct use of water to achieve evaporative cooling is problematic in warm, humid climates, an indirect method of achieving the same is provided by the evapotranspiration potential of vegetation. Unlike open water bodies, which simply add moisture to an already moist tropical environment, trees utilise water *and* solar energy to evapotranspire energy, thus reducing solar radiation's impact on human comfort. While photosynthesis is partly responsible for the reduction in radiation beneath a tree, the greater part of thermal comfort under a tree results from the evaporation of water from leaves exposed to the sun (see Givoni, 1991). This will add to higher humidity, but, as the discussions on thermal comfort show, thermal discomfort is controlled greatly by mean radiant temperature (MRT) rather than the humidity levels. Thus planted areas offer excellent opportunities to promote evaporative cooling in the tropics.

Givoni (1991: 289) suggested the following as the main design details affecting planted areas' contribution to the improvement of indoor and outdoor comfort:

1   width of planted area around the building;
2   type of plants: trees, shrubs, lawn, vines climbing on walls, and so on;
3   size and shape of trees and shrubs;
4   orientation of plants with respect to the building.

The larger the width of planted areas the greater the cooling potential in warm, humid areas. The influence of type of vegetation is not so much on the evaporation potential but on its wind impeding effects. Isolated, tall-stemmed vegetation poses little threat to local level wind flow (in fact it might help concentrate air flow below the tree canopy, thus improving ventilation near the ground; see Givoni, 1991), but bushes and short vegetation could potentially impede air flow around buildings.

CONDUCTIVE COOLING BY EARTH ARCHITECTURE

The absence of temperature variations across the year makes the difference between air and ground temperatures relatively small in warm, humid areas. Added to that, higher rainfall keeps the ground water table high most of the year. These factors together have contributed to the near absence of the practice of building below ground level. Yet some possibilities exist, especially in rolling terrain where earth architecture might provide thermal relief in the oppressive wet tropics.

Earth is a good insulator and therefore has been a valuable design tool in hot and dry climates that witness extreme diurnal temperature swings. Earth berms help reduce the magnitude of temperature variations in such climates, thereby making the indoors more tolerable. However, in the hot and humid tropics, earth embankments will have to act primarily as a solar shading apparatus rather than as insulation media. This would also mean that such shielding of sun might be problematic at night when outside conditions are quite tolerable.

Solar radiation prevention is the most important contributor to daytime comfort in underground spaces. But, unless these spaces are allowed to benefit from the cool outside at night, their night-time comfort would rapidly deteriorate. Thus conductive cooling in warm, humid areas is a mixture of radiation prevention (strategy no. 1) and convective cooling (strategy no. 2).

Case study: Zero carbon non-domestic buildings in warm, humid climates: MAS Intimates Thurulie, Thulhiriya, Sri Lanka

This case study details a clothing factory built in the warm, humid western region of Sri Lanka that recently obtained the US Green Building Council's LEED certification, Platinum Grade, for its sustainability credentials. A 10,000-square-metre facility producing lingerie for export to the UK, it employs 1,300 people (over 1,100 of whom are sewing machine operators) and was built at a cost of USD 2.66 million in mid-2008.

To realise a sustainable design, the design team applied a three-point philosophy of respect for the site, respect for users and respect for ecosystems. These three aims, complementing the functional and commercial requirements for the project, served as selection criteria for all materials and systems used in the building (Wentz, 2009). The design is inspired by traditional Sri Lankan architecture, built partially on stilts, with courtyards, amid lush greenery.

The building is powered by carbon-neutral energy sources (hydro and solar PV) and uses half the water of comparable factories, even though the grounds are a veritable garden. The facility incorporates an anaerobic digestion system for sewage treatment.

Figure 5.8 An LEED Platinum certified building in warm, humid climates, with cool and green roofs and shading by vegetation

The building is designed for efficient production, a comfortable atmosphere and low energy consumption. Meeting these three criteria in the tropics means mastering above all one thing – cooling. Cooling is achieved at the plant primarily by passive design and secondarily by active systems. Passive design measures include the orientation and massing of building volumes, controlled fenestration and ventilation, shading of the building and its surroundings, and thermal mass and solar reflectivity of the façades and roofs.

The high angle of the sun during most of the year makes the south façade the easiest to shade and the east and west façades the most difficult. Thus the main building volumes, the production spaces, are aligned on an east–west axis, the north and south façades being the largest. This orientation makes it easier to block direct solar radiation.

The massing of the building volumes and the positioning and sizing of windows permit daylight to enter as natural illumination without causing substantial heat gain. Horizontal shading intercepts the northern sun at mid-year and the southern sun later in the year.

Thermal roof load, the largest contributor to heat gain and indoor discomfort in the tropics, is controlled by a combination of green roofs, photovoltaic roofs, and cool roofs. Green roofs cover 1,757 square metres of the building. The photovoltaic roof covers 200 square metres of the roof, while the cool roof is a lightweight metal roof assembly over the long-span production halls. The white metal roof reflects nearly 80 per cent of the solar energy that reaches the roof.

Another passive means of keeping the building cool is to cool the microclimate, or reduce the ambient heat around the building. The heat island effect around the building is controlled by shading, covering parking areas, lighter, reflective paving around the building instead of dark, heat-absorbent paving, and shading the courtyards between the building volumes. The combination of the many passive cooling measures reduces the thermal load to a level that can be handled by low carbon power sources.

A further reduction in cooling energy need is achieved by evaporative cooling units. These units draw in fresh air, filter it, and add moisture to lower the dry bulb temperature. The air is distributed through a balanced system of ducts and fed into the spaces, which remain under positive static pressure. Indoor air is not recirculated, but extracted by suitably sized exhaust fans to ensure effective moisture and heat removal. The air exchange rate is about 40 air changes per hour and is perceptible (about 0.8 m/s). This will lead to the extension of the thermal comfort zone by about 2.7°C (Halwatura and Jayasinghe, 2008).

The combination of microclimate improvement, cool, green and solar PV roofs and a low energy evaporative cooling system leads to an indoor dry bulb temperature that is up to 3°C cooler than the outdoors and the indoor relative humidity about 10 per cent higher than the outdoors (Wentz, 2009). In order to further enhance comfort, building users are encouraged to wear appropriate clothing (short-sleeve shirts, with many workers going barefoot; Wentz, 2009). The combination of cool dressing, activity at low metabolic rates, and air movement makes the plant a comfortable working environment. The maximum observed temperature on the ground floor of the building is 29.5°C, which is acceptable, because the indoor air velocity of 0.8 m/s keeps the environment within the extended comfort zone.

Energy consumption for lighting was reduced by maximising daylighting and by using well-designed systems with efficient lamps. Offices, cafeteria, lounge, reception area, meeting rooms and boardrooms are normally illuminated by daylight only. The glare-free illumination is usually adequate even on rainy days. Daylight is adequate roughly from 6 a.m. to 6 p.m., which easily covers the normal operating hours of the plant. An added benefit of forgoing artificial lighting is reduced heat gain within the building.

Strategies to reduce embodied energy (energy expended to process and transport materials) include compressed stabilised-earth block walls locally manufactured. Beside the local production, compressed walls require no plaster finish, further reducing the material needed to finish the wall surfaces. Bamboo is used for window blinds and various forms of sun screen. Non-hazardous finishes and materials are used throughout the building, ensuring good indoor air quality, which is enhanced by high air exchange rates.

5.3.3 Hot, dry climates

In the hot, arid climates of West and Central Asia, North Africa and South America the key to low/zero carbon buildings is to restrict solar gain as in the warm, humid case, but unlike in the latter simultaneously reduce infiltration. While evaporative cooling will offer an advantage, the lack of water availability poses serious problems. Instead, the large daily (diurnal) temperature variations provide the opportunity to cool interiors radiantly.

- restrict solar gain;
- restrict conduction;
- restrict infiltration;
- promote radiant cooling.

The first two strategies are detailed in 5.3.2 above; the third (air tightness to reduce infiltration) is extensively dealt with under the Passivhaus approach (see 5.3.1 above). The promotion of radiant cooling involves the careful manipulation of thermal mass, ventilation and shading to reduce surface temperatures, which can in turn promote heat loss by radiation.

Radiant cooling in hot, dry climates

Low equivalent sky temperatures in arid regions have been used for radiative cooling of heavy roofs in traditional architecture (Rosenlund, 2000). The approach is to insulate the roofs for solar protection during the day *and* reduced heat losses in cold seasons.

Rosenlund (2000) presented the following strategies as appropriate for hot, arid climates:

- Layout: compact urban plan; narrow streets that could be covered or lined by arcades create shade and act as cool ponds; protection from air borne sand and dust; highly reflective surfaces; solar access during colder seasons.
- Building form: compact building form to minimise conduction; courtyard structures to create intermediate zones of changing local climates; north–south orientation of the main façades; solar protection is important especially towards the west.
- Other approaches: fountains and vegetation in courtyards; radiative cooling towards the clear sky especially from surfaces with a high sky view factor, such as a roof; heavy materials to moderate the internal temperature swings; night cooling.

Figure 5.9 An example of appropriate design in hot, dry climates

Photo credit: Hans Rosenlund.

Figure 5.10 Radiant cooling in hot, dry climates by manipulating building thermal mass and shading

Photo credit: Hans Rosenlund.

Note: Aqaba Residential Energy Efficiency (AREE) building, the first of its kind in Jordan, built in 2007. The concept includes thermal insulation and mass, shading, night ventilation, earth cooling, a roof garden, grey water recycling for irrigation, low energy lighting and appliances, and solar cooling. Architect: Florentine Visser. Sustainability evaluation: Hans Rosenlund, CEC Design.

5.4 Problems needing urgent action

Two aspects of a changing climate pose a serious threat to low carbon buildings: overheating in all climates and the need to clarify thermal comfort standards to reflect climate change.

5.4.1 Overheating

The changing climate and local warming in urban areas due to haphazard urban growth and efficient heat trapping by building geometry are twin realities that contemporary buildings, especially those in urban areas, must face. Many of the climate-specific design strategies presented in 5.3 above assume little change in the background climate, an assumption increasingly difficult to sustain. Added to this, an over-reliance on active energy to heat or cool contemporary buildings has diminished the adaptive capacity of buildings to deal with local and global warming. This is in sharp contrast to traditional architecture that made use of shading, ventilation, thermal mass and other passive technologies to avoid or lessen the effects of overheating and create a liveable, if not always a comfortable, indoor environment (Nicol, 2009).

5.4.2 Thermal comfort standards

Even as the building envelope's ability to deal with a changing climate is diminished, our expectation for thermal comfort has dramatically increased. The over-reliance on heating, ventilating and air conditioning (HVAC) equipment to enhance indoor comfort in all climates in all buildings throughout the year cannot be met without serious carbon consequences. The current approach to thermal comfort is characterised by the 'commodification' of comfort (see Fanger, 1970), which is defined in ways that only the HVAC industry could fulfil (Nicol, 2009).

Comfort expectations are so deeply imprinted in modern lives that strategies for low carbon alternatives still remain mechanically based, something that will probably lead to an 'efficient unsustainability' as Rees (2009) termed it. What were seen as 'luxuries' for the few (such as air conditioning) end up being integrated into everyone's lives in such a way as to eventually become essential (Shove, 2003).

The problem of overheating as well as the approach to thermal comfort is perhaps best exemplified by the modern use of air conditioners. It is estimated that as much as 70 per cent of all delivered energy to air conditioned buildings in the United States is used to power the air conditioning itself (Nicol, 2009). A dangerous positive feedback has been created. Yet a simple return to traditional passive and low carbon solutions to overheating on any significant scale is almost certainly not possible.

Shove *et al.* (2008) comment that, in global terms, the energy cost of maintaining standardised 'comfort' conditions in buildings and in outdoor environments around the world is ultimately unsustainable. The expectation and the reality of hotter, more extreme or more extremely varied outdoor climates are already significant for the definition and provision of comfort in what will also have to be a lower carbon society (Shove *et al.*, 2008).

New ways of thinking about comfort will be needed to achieve a low carbon but 'comfortable' building stock. These include:

1   A realisation that meanings and definitions of comfort are not set in stone. Expectations may change in ways that exacerbate or reduce problems of climate change, but the crucial point is that they will change and that neither direction is a foregone conclusion.
2   Historical differences in the meaning, framing and material management of offices, homes, churches, cars, gardens and hospitals result in a plethora of distinctly different opportunities for intervention, including a range of adaptive strategies associated with hybrid formulations like the home office, the car as living room, or the patio as dining area.
3   'Demand' and 'supply' are connected, and the future of the indoor environment is in no small measure bound up with the ambitions, discourses and problem definitions of powerful providers. The key point here is that manufacturers, designers, scientists and policy makers are the sometimes unwitting (sometimes deliberate) purveyors of ideas and images as well as standards, regulations and technologies of indoor climate control. Again, the practical implication of this insight is clear: since all these actors and more are actively involved in making and shaping the future of comfort, all represent possible, variously effective points of influence (Shove *et al.*, 2008: 307).

5.5 Drivers and barriers to LZC in the new build

Given the legislative mandate for a low carbon economy, the UK provides an interesting case study of drivers of, as well as barriers to, low/zero carbon buildings. Ward (2008) states that a great deal more needs to be done to stabilise and reduce the amount of energy consumed by the building sector if the UK is to meet its 60 per cent reduction target by 2050. In the case of LZC houses, Ward (2008) identified the main drivers for LZC in the UK to be:

- increased numbers of households: 3 million new homes are projected by 2020;
- a government commitment to low-carbon homes;
- the introduction of the Code for Sustainable Homes;
- the introduction of home information packs;
- a social change in energy attitude and behaviour;
- the security of supply to vulnerable groups.

Counteracting the drivers, a number of barriers have also been identified which could act against achieving energy reductions. These are:

- the oldest housing stock in Europe, being replaced at about 1 per cent per year;
- an increased use of electrical equipment;
- an increased use of conservatories as extra heated living space;
- the relaxation in planning permission for extensions, including conservatories;
- an ageing population living in older properties not being able to afford improvements;
- the increased use of condensing boilers and combi-boilers, reducing the ability to use solar or other renewable energy that needs storage for hot water;
- tighter building regulations may mean more overheating and an increased use of domestic air conditioning (a 415 per cent sales increase recorded in 2006);
- limited information on the reliability of emerging technologies for energy services;
- a lack of user information to help with the use and understanding of domestic heating systems, including renewable systems;
- an increase in disposable income.

References

All websites were last accessed on 30 November 2011.

Acharya, P.K. 1979. *Indian Architecture According to Manasara-Silpasastra*. Patna: Indian India.
Chappells, H. and Shove, E. 2003. The environment and the home. Draft paper for Environment and Human Behaviour Seminar, 23 June, Policy Studies Institute.
Correa, C. 1989. *The New Landscape: Urbanization in the Third World*. London: Butterworth Architecture.
DCLG (Department of Communities and Local Government). 2006. *Code for Sustainable Homes: A Step-Change in Sustainable Home Building Practice*. London: DCLG. Available at: http://www.planningportal.gov.uk/uploads/code_for_sust_homes.pdf.
DECC (Department of Energy and Climate Change). 2009. *Digest of United Kingdom Energy Statistics 2009*. London: Stationery Office.
DECC. 2011. *CO$_2$ Emissions by National Communication Sectors for the Period 1990–2010 (Provisional)*. London: DECC. Available at: http://www.decc.gov.uk/en/content/cms/statistics/climate_stats/data/data.aspx.
Emmanuel, R. 2005. *An Urban Approach to Climate Sensitive Design: Strategies for the Tropics*. London: Spon Press.

Emmanuel, R., Kumar, B., Roderick, Y. and McEwan, D. forthcoming. A universal climate-based energy and thermal expectation index: initial development and tests.

Fanger, P.O. 1970. *Thermal Comfort: Analysis and Applications in Environmental Engineering.* New York: McGraw-Hill.

Federer, C.A. 1971. Effects of trees in modifying urban microclimate. In *Trees and Forests in an Urbanizing Environment*, ed. S. Little and J.H. Noyes, pp. 23–28. Amherst: University of Massachusetts Cooperative Extension Service.

Feist, W. 2008. *What Is a Passive House?* Darmstadt: Passive House Institute. Available at: http://www.passivhaustagung.de/Passive_House_E/Passive_House_in_short.html.

Givoni, B. 1989. *Urban Design in Different Climates*, Technical Note WMO/TD-346, WCAP-10. Geneva: World Meteorological Organization.

Givoni, B. 1991. Impact of planted areas on urban environmental quality: a review. *Atmospheric Environment B*, **25**, pp. 289–299.

Givoni, B. 1994. Urban design for hot, humid regions. *Renewable Energy*, **5**, pp. 1047–1053.

Grant, Z., Garrod, A., Myers, K. and Nolan, M. 2009. *Towards a Low Carbon Built Environment: A Roadmap for Action.* London: Royal Institution of Chartered Surveyors (RICS).

Halwatura, R.U. and Jayasinghe, M.T.R. 2008. Thermal performance of insulated roof slabs in tropical climates. *Energy and Buildings*, **40**, pp. 1153–1160.

Heisler, G.B. 1974. Trees and human comfort in urban areas. *Journal of Forestry*, **72**, pp. 466–469.

IES. 2011. *Integrated Environmental Solutions Virtual Environment (IES VE).* Available at: http://www.iesve.com/software.

Janda, K.B. 2009. Worldwide status of energy standards for buildings: a 2009 update. In *Proceedings of the European Council for an Energy Efficient Economy (ECEEE) Summer Study, Cote d'Azur, France, June 1–6, 2009*, pp. 485–491. Available at: www.eci.ox.ac.uk/publications/downloads/janda09worldwidestatus.pdf.

Kelly, M.J. 2009. Retrofitting the existing UK building stock. *Building Research and Information*, **37**, pp. 196–200.

Koenigsberger, O.H., Ingersoll, T.G., Mayhew, A. and Szokolay, S.V. 1974. *Manual of Tropical Housing and Building*, part I: *Climatic Design.* London: Longman.

Köppen, W. 1923. *Die Klimate der Erde.* Berlin: Walter de Gruyter.

Kramer, P.J. and Kozlowski, T.T. 1960. *Physiology of Trees.* New York: McGraw-Hill.

Leonard, R.E. 1972. Making our lives more pleasant – plants as climate changers. In *Landscape for Living*, pp. 5–9. Washington, DC: USDA.

Levine, M., Ürge-Vorsatz, D., Blok, K., Geng, L., Harvey, D., Lang, S., Levermore, G., Mongameli Mehlwana, A., Mirasgedis, S., Novikova, A., Rilling, J. and Yoshino, H. 2007. Residential and commercial buildings. In *Climate Change 2007: Mitigation: Contribution of Working Group III to the Fourth Assessment Report of the Intergovernmental Panel on Climate Change*, ed. B. Metz, O.R. Davidson, P.R. Bosch, R. Dave and L.A. Meyer. Cambridge: Cambridge University Press.

Lohmann, U., Sausen, R., Bengtsson, L., Cubasch, U., Perlwitz, J. and Roeckner, E. 1993. The Köppen climate classification as a diagnostic tool for general circulation models. *Climate Research*, **3**, pp. 177–193.

Mather, J.R. 1974. *Climatology: Fundamentals and Applications.* New York: McGraw-Hill.

Miller, R.W. 1988. *Urban Forestry: Planning and Managing Urban Green Spaces.* Englewood Cliffs, NJ: Prentice Hall.

Monahan, J. and Powell, J.C. 2011. A comparison of the energy and carbon implications of new systems of energy provision in new build housing in the UK. *Energy Policy*, **39**, pp. 290–298.

Monteith, J.L. 1973. *Principles of Environmental Physics.* London: Edward Arnold.

Morgan, M.H. (trans.). 1960. *Vitruvius: The Ten Books on Architecture.* New York: Dover.

Myrup, L.O. 1969. A numerical model of the urban heat island. *Journal of Applied Meteorology*, **32**, pp. 896–907.

NBT (Natural Building Technologies). 2009. *Passivhaus: Timber Frame Systems.* Oakley, Bucks: Natural Building Technologies. Available at: http://www.natural-building.co.uk/PDF/Pavatex/090216_Technical_Manual_PASSIVHAUS.pdf.

Nicol, F. 2009. Cooling in a low carbon world. *Building Research and Information*, **37**, pp. 345–347.

Nieuwolt, S. 1986. Design for climate in hot, humid cities. In *Urban Climatology and Its Applications with Special Regard to Tropical Areas*, Proceedings of a Technical Conference organised by the World Meteorological Organization, Mexico City, 26–30 November 1984, pp. 514–534. Geneva: World Meteorological Organization.

Oke, T.R. 1987. *Boundary Layer Climates*. London: Methuen.

Oke, T.R. 1989. The micrometeorology of the urban forest. *Philosophical Transactions of the Royal Society of London B*, **324**, pp. 335–350.

Parker, J. 1983. Do energy conserving landscapes work? *Landscape Architecture*, **72**, pp. 89–90.

RAE (Royal Academy of Engineering). 2010. *Engineering a Low Carbon Built Environment*. London: Royal Academy of Engineering. Available at: www.raeng.org.uk.

Rees, W.E. 2009. The ecological crisis and self-delusion: implications for the building sector. *Building Research and Information*, **37**, pp. 300–311.

Rosenlund, H. 2000. Climatic design of buildings using passive techniques. *Building Issues*, **10** (1), pp. 1–26.

Rubel, F. and Kottek, M. 2010. Observed and projected climate shifts 1901–2100 depicted by world maps of the Köppen–Geiger climate classification. *Meteorologische Zeitschrift*, **19**, pp. 135–141.

Shove, E. 2003. *Comfort, Cleanliness and Convenience: The Social Organization of Normality*. Berg: Oxford.

Shove, E., Chappells, H., Lutzenhiser, L. and Hackett, B. 2008. Comfort in a low carbon society. *Building Research and Information*, **36**, pp. 307–311.

Szokolay, S.V. 1992. *Architecture and Climate Change*. Red Hill, ACT: Royal Australian Institute of Architects, Education Division.

Ürge-Vorsatz, D., Harvey, L.D.D., Mirasgedis, S. and Levine, M.D. 2007. Mitigating CO_2 emissions from energy use in the world's buildings. *Building Research and Information*, **35**, pp. 379–398.

Wang, L., Gwilliam, J. and Jones, P. 2009. Case study of zero energy house design in UK. *Energy and Buildings*, **41**, pp. 1215–1222.

Ward, I.C. 2008. What are the energy and power consumption patterns of different types of built environment? *Energy Policy*, **36**, pp. 4622–4629.

Wentz, D. 2009. *MAS Intimates Thurulie: Clothing Factory in Sri Lanka*. Zurich: Holcim Foundation. Available at: www.holcimfoundation.org/mas.

Yannas, S. 1990. *Passive Solar Heating and Energy Efficient Housing Design*, Proceedings of the 2nd European Conference on Architecture, pp. 548–556. Dordrecht: Kluwer.

Yannas, S. 1994. *Solar Energy and Housing Design*, vol. 1: *Principles, Objectives, Design Guidelines*. London: Architectural Association Publications.

Yannas, S. 1996. Energy indices and performance targets for housing design. *Energy and Buildings*, **23**, pp. 237–249.

6 Carbon management in the existing stock

Admittedly the manoeuvrability of existing buildings to make them low/zero carbon is limited. Nevertheless, existing building stocks present the greatest challenge to carbon management in the built environment. For example in the UK, even with its legislative mandate to decarbonise the economy, nearly 87 per cent of the buildings that will be occupied in 2050 (the year in which UK law requires the economy to have 60 per cent less carbon than it did in 1990) have already been built. New construction in Europe and the United States annually amounts to around 1 per cent of the existing building stock; however, twice that many structures are renovated every year (Butler, 2008). The wastage of energy in existing buildings not only is a carbon problem but also seems to contribute to ill health and even mortality risk (Wilkinson *et al.*, 2007). Thus the opportunities and potential presented by the existing stock are enormous. At the same time, inaction at the renovation stage of buildings locks in carbon wasteful practices for years to come.

The carbon performance of existing buildings varies with type of construction, age of buildings, technologies used and, in the case of housing, density of dwelling units. The increasing stringency of building regulations, coupled with greater awareness of energy performance issues, is slowly leading to better carbon performance in new buildings, but there is a considerable distance to go before significant reductions in carbon are achieved across the whole of existing building stock. Table 6.1, from the Scottish House Condition Survey (Walker *et al.*, 2010), provides a typical example. The newer buildings are more carbon efficient, and the more dense dwelling arrangements (such as tenements and flats) are more than twice as energy efficient as newer detached buildings. Across the entire stock the variation (i.e. ratio between the worst and

Table 6.1 Mean CO_2 emissions in dwellings in Scotland

Age of dwelling	Fraction of national total (%)	Emission by type of dwelling (tonnes/yr)					Emission by age of dwelling (tonnes/yr)
		Detached	*Semi-detached*	*Terraced*	*Tenement*	*Other flats*	
Pre-1919	19	17.5	11.2	9.5	5.2	8.8	9.9
1919–44	14	15.7	8.1	6.1	4.2	4.4	7.0
1945–64	22	12.4	6.7	5.3	3.9	4.3	5.7
1965–82	23	9.2	6.0	5.0	4.2	3.9	5.9
Post-1982	22	7.5	4.5	4.0	3.3	3.1	5.2
All	100	11.0	7.0	5.6	4.2	4.8	6.6

Source: Based on data from Walker *et al.*, 2010.

the best performing dwelling) in carbon performance is a staggering 5.5. A key to the scale of the problem is indicated by the stock sizes according to age. Nearly a third of the dwellings are over 65 years old (and emit more carbon than the stock average). Thus a large portion of the task of decarbonising the existing housing stock still remains to be done.

Low/zero carbon (LZC) in existing buildings can be achieved by:

- climate-specific solutions;
- universally applicable improvements to decarbonise buildings.

Climate-specific solutions attempt to work with the background climate to achieve energy efficiency and therefore carbon reduction. In cooler climates, an LZC approach to existing buildings ought to increase the envelope's ability to resist *heat loss*, whereas in warmer climates the opposite needs to be true: reduce *heat gain*. Universally, vegetation based solutions can enhance the climatic suitability of existing stock, irrespective of local climate. Similarly, energy efficiency in building appliances (especially lighting) can reduce the carbon footprint of all buildings, irrespective of the climatic contexts. The generation of clean (renewable) energy at site is a further option to reduce the carbon footprint of existing buildings in all climates.

6.1 Retrofitting buildings for low/zero carbon – climate-specific solutions

6.1.1 Retrofit in temperate climates

In cooler climates, the principal LZC approach to existing buildings is to reduce heat loss. This can be achieved by adding thermal barriers (insulation) to roofs, walls and/or floors and increasing the air tightness of buildings (mainly through windows and openings) to reduce heated indoor air from escaping into the outside. Furthermore, efficient mechanical heating systems will bring in a significant reduction in the energy (and carbon) needed to keep the indoors comfortably warm.

Insulation

Table 6.2 summarises the costs and benefits of common insulation options which are applicable to existing properties in cooler climates. The figures are based on a gas centrally heated semi-detached house in the UK.

Table 6.2 Insulation options for existing buildings in a cooler climate (the UK)

Measure	Installed cost	Installed payback	Annual savings (£/yr)	CO2 saving (t/yr)
External wall insulation	£4,500	12 years	£380	2.60
Internal wall insulation	—	—	£130–£360	2.40
Loft insulation (from none to 270mm)	£500	3 years	£155	1.00
Loft insulation (increase from 50mm existing to 270mm)	£500	11 years	£45	0.35
Floor insulation	—	—	£40	0.25
Hot water tank jacket	N/A	N/A	£30	0.20
Draught-proofing	£200	8 years	£25	0.15
Filling gaps between floor and skirting board	N/A	N/A	£20	0.13
Primary pipe insulation	N/A	N/A	£10	0.07

Source: EST, 2008b.

Note: All figures are approximate and based on a gas centrally heated semi-detached house in the UK.

Heat loss through all building envelope elements (walls, roofs and floors) could be reduced by insulation. In terms of wall insulation, the option for existing buildings is to apply insulation to either the exterior (more effective, since heat is retained in the building element itself) or the interior. External wall insulation is generally not feasible for high rise properties and may not be permitted for buildings with significant historic value; however, it is one of the most effective strategies to insulate buildings with solid walls. Adding internal wall insulation is often the only effective measure to improve energy efficiency in existing buildings. Internal insulation comes in two basic forms: a lining or fill fitted between the batons holding the plasterboard or added as an additional layer in itself; and solid boards or a dry liner that are attached to the existing wall. The former requires significant building work and is generally likely to be attractive during major refurbishment, whilst the latter generally does not require expert installation and can be added during general redecoration. Care needs to be taken to ensure that the properties of the insulation and the wall are compatible, as a permeable wall insulated with an impermeable insulation liner will produce damp problems and damage the wall. Adding internal insulation may require the removal of internal fixtures and electrical installations; therefore special care needs to be taken with historic buildings. In some cases it should be permissible to use a thin insulative lining to replace the existing wall lining, whereas in others it may be more useful to consider a wider refurbishment including the replacement of plasterboard.

Loft and floor insulation have lower impacts but are easier and less intrusive to install than internal wall insulation, and floor insulation may be particularly attractive to residents of tenements where the traditional wooden floorboards remain exposed. Table 6.2 also shows that simple measures such as filling gaps and insulating pipes and water tanks can produce smaller but appreciable reductions in carbon and enhanced energy efficiency.

Windows and glazing

Although only around 10 per cent of a dwelling's heat losses are through its windows, replacement of faulty windows provides a visible and often audible evidence of 'improvement'. The draft and heat loss through ill-fitting windows often leads to visible condensation and excessive noise ingress, both of which can be cured by tight-fitting frames with multi-layered glazing.

Installing energy efficient double glazing can reduce heat losses through windows by half and save around 680 kilograms of CO_2 a year (EST, 2012). Except in extremely cold climates, triple glazing is usually unnecessary; the relatively minor improvements in energy efficiency over double glazing tend not to be cost-effective, but it has the added value of being more effective for reducing noise pollution.

SECONDARY GLAZING

Installing secondary glazing is a cheaper option for reducing heat loss and sound ingress through windows, and provides an alternative to double glazing for buildings where window replacement is impractical (such as high rises or historically significant buildings). A wide variety of types exists, and not all are suitable for all properties. They vary from adding a second internal window to adding layers on the back of existing glazing inside the original frame or adding internal shutters. The former is a more effective solution, particularly for blocking sound, but is more expensive and may intrude into the room. The latter can be less intrusive, and can be custom designed to meet specific constraints in existing buildings, but keeping at least a 150 millimetre gap between the primary and secondary glazing is recommended for there to be an appreciable impact on sound.

In general the system should be chosen to be as invisible as possible from the outside: any bars on the internal window should match those on the original; ideally the method of opening should match that of the existing window; and the glazing and any coatings that may be applied to it should be chosen to avoid the problem of 'double reflection'. The chosen system should not prevent the easy opening of the existing window, both for safety reasons in the event of needing to use the window for escape in an emergency, and so as not to restrict ventilation. Some systems are designed to be removed during the summer, and so the panes require safe storage to avoid potentially serious and costly accidents. As an alternative many householders opt for perspex inner secondary glazing frames, as these are cheaper, lighter and easier to remove and replace. Internal shutters may be more effective in buildings not occupied during the day, as they need to be closed to reduce heat losses in winter and heat gains in summer, but they are also useful if only used as an additional layer of insulation at night. However, they may be incompatible with curtains (which have a similar effect), and so choosing whether or not to install them should be based on their aesthetics as well as their practicalities.

DRAUGHT-PROOFING

Draught-proofing is a cheap and cost-effective measure to increase the air tightness of windows and can have the additional benefit of eliminating rattling in loose-fitting windows. A wide range of options exists, some of which are more visible, costly and difficult to install than others, which should mean that even residents of listed buildings will have several to choose from. Care should be taken to choose an option that does not block access to the opening mechanism for the window or risk jamming it shut, particularly in the case of sash windows (Changeworks, 2008a, 2008b).

Heating systems and controls

The main factor that will dictate the choice of system is whether or not the property has a mains gas supply. It is possible to use LPG, paraffin or wood-burning stoves as alternatives, but LPG and paraffin stoves can produce moisture problems, and wood burners, although more sustainable, require storage space for fuel.

For buildings with radiators further gains can be made by simple additional improvements. An aluminium foil sheet can be added behind the radiators to reflect heat back into the occupied areas of the room, and small shelves can be added to help direct heat into the room space (this may not be permissible for historic properties).

Improving the level of control householders have over their heating can also produce notable reductions in energy consumption, and there is some evidence that occupants with more control do heat their homes more optimally (Roaf *et al.*, 2008). For gas fuelled systems the highest levels of control can be achieved when all of the following are installed:

- electronic (ideally fully programmable) timer;
- room thermostat;
- thermostatic radiator valves (TRVs); and
- thermostat on the hot water tank (or on the boiler in the case of combination boilers).

All these can be retrofitted to most gas heating systems and will have very little or no impact on the appearance or space availability in existing properties. If the system is so old that fitting additional controls is problematic it is likely to be worth investing in a complete replacement (Roaf, Baker and Peacock, 2008).

Table 6.3 Options to reduce emissions in hard to treat existing buildings in temperate climates

Property type	Primary measure(s)	Secondary measure(s)
Solid wall	Prioritise cost-efficient sets of interventions that are appropriate to the individual requirements of the solid wall house type.	Eliminate cold bridging. Eliminate infiltration. Experiment with new insulation types. Reduce energy use through options applicable to all or most hard to treat properties.
Tenement	Aim to have all blocks covered by factoring agreements that include carbon factoring; use these to ensure all blocks are brought up to recommended energy efficiency standards. Promote and subsidise secondary glazing and draught-proofing. Develop or amend factoring legislation to include carbon factoring.	Use factoring schemes to offer energy audits for individual flats and use these to promote improvements such as heating system upgrades, low energy lights, energy efficient appliances and behavioural changes. Promote and subsidise the construction of draught lobbies and the installation of under-floor insulation. Design and install micro-CHP schemes specifically for groups of tenements.
High rise	Use reliable and replicable protocols for evaluating the cost efficiency of high rise investments. Externally insulate. Improve the efficiency of lifts and water pumping systems. Use combined heat and power (CHP) systems. Include management issues in solutions.	Consider connecting CHP and building energy services to other local buildings. Use intelligent procurement to find ways of reducing over-cladding costs. Lifecycle cost all decisions.
Timber frame	Promote the uptake of external and internal cladding suitable for timber frame walls.	Reduce energy use through options applicable to all or most hard to treat properties.

Flat roof	Install warm deck flat roofs on all suitable properties.	Where installing a warm deck is not possible, promote the installation of internal cladding on ceilings.
Mansard roof	Install blown fibre insulation behind the tiles on the lower section (note that this needs to be carried out in dry weather). Treat upper sections as for normal sloping roofs where access allows.	Install internal cladding on the walls and ceilings of rooms contained within the mansard.
Park homes and residential mobile homes	Install internal wall, ceiling and floor insulation, external cladding or render, and the sealing of leaks around doors and windows. Legislate to ensure all landlords of rented homes bring insulation up to the recommended levels. Revise product standards to improve minimum energy efficiency standards.	Installing double glazing will benefit some homes, but has a long payback period. For homes with flat roofs, adding an insulated sloping roof may be an option, but payback periods suggest that offering subsidies is not cost-effective. Promote solar thermal and micro renewables, particularly to homes not on mains gas or electricity supplies.
All hard to treat	Target all hard to treat properties with information on secondary glazing and draught-proofing. For properties with gas central heating, promote the replacement of old boilers with newer combi or condensing models; encourage annual servicing and installation of improved controls (TRVs, thermostats, programmable timers). For properties without a mains gas supply promote the upgrading of heating systems to the latest, most energy efficient designs. Target all properties with information on the savings that can be made from using low energy light bulbs and replacing old appliances with A rated and above. Subsidise or incentivise the above, particularly secondary glazing and draught-proofing, and prioritise older properties and households on low incomes.	Install smart meters in all homes. Develop and promote community CHP schemes. Encourage the uptake of micro-CHP and renewables, and consider revising planning and conservation legislation where feasible.

Source: Roaf, Baker and Peacock, 2008.

Additional issues facing non-domestic buildings

While the basic approach to carbon reduction in non-domestic buildings in temperate climates is similar to the discussions above (insulation, windows and glazing, heating and controls) a key difference is the internal thermal loads in large non-domestic buildings, which make them behave as cooling dominated buildings even in temperate climates. The internal activity of non-domestic buildings is crucial, and the efficiency of small power and lighting should be improved before any other measures are taken (Jenkins, Banfill and Pellegrini-Massini, 2009). If care is not taken, increased insulation and air tightness resulting from the strategies specified above could lead to an overheating problem, especially under a climate change scenario. To counter this threat and the associated CO_2 emissions penalty (i.e. the increased need for heating *and* cooling) effective building design and correct management of IT equipment and lighting are key (Jenkins *et al.*, 2009).

Jenkins *et al.* (2009) highlighted the following key issues facing specific non-domestic buildings in temperate climates:

1   Greater care is needed to avoid the overheating problems (especially in schools and offices).
2   Reducing lighting energy consumption in non-domestic buildings through improving technologies provides an easy option for reducing carbon (halogen spot-lighting in shops has a negative effect).
3   Open fronted display refrigerators in supermarkets are a major source of CO_2 emissions and contribute to heating energy consumption through local cooling.
4   On-site energy generation can achieve significant savings only if very large systems are installed, and these are difficult or impossible to justify economically. Integration with the existing network infrastructure may help, but the goal should be an overall reduction in the CO_2 intensity of delivered energy.
5   Capital and whole lifecycle costs of technologies needed for large emissions reductions (especially beyond 50 per cent) in non-domestic buildings are high, and there is not sufficient attraction for owners, landlords and managers of existing properties to employ all of the technologies identified above.
6   The goal of 'net zero' carbon non-domestic buildings will not be achieved, by any definition, without dramatically reducing the energy consumption of small power and lighting, since few existing buildings will be able to satisfy their electrical energy demand through on-site generation of energy using renewable sources, such as photovoltaics (PV), wind and combined heat and power (CHP) supplies (Jenkins *et al.*, 2009).

Given the internal load domination in non-domestic buildings and the limited possibilities in improving fabric conditions and/or heating and control system enhancements, Jenkins *et al.* (2009) suggested the following options for carbon reduction in existing non-domestic buildings:

1   *Lighting and small equipment:*

   a   IT energy management (including overnight switching off of non-essential computer servers);
   b   cholesteric liquid crystal display (LCD) monitors replace cathode ray tube (CRT) monitors;
   c   reduced computer usage with more efficient process;
   d   multifunction machines for all printing, copying and scanning;
   e   light emitting diode (LED) lighting (150 lumens/Watt) replacing fluorescent (such as T12) tubes (70 lumens/W).

2   *Fabric:*

   a   external insulation of expanded polystyrene (EPS) with concrete render;
   b   EPS also used for floor (100mm) and roof (200mm), replacing existing mineral wool;
   c   triple glazed argon windows (U-value = 0.78W/m²K), with low-e coating, replacing existing double glazing;
   d   infiltration reduced from 1 air change per hour (ach) to 0.5 ach.

3   *Heating and controls:*

   a   condensing boiler replaces non-condensing boiler;
   b   mechanical ventilation heat recovery (MVHR);

c   adaptive comfort approach to cooling;
d   reduction in internal gains (via efficient small appliances and lighting).

Table 6.4 shows the likely carbon and energy consumption reductions possible for a variety of non-domestic buildings in the UK if the above strategies are utilised.

6.1.2 Retrofit in warm climates

The key need in warm climates is to reduce heat gain. This can be achieved by:

*   external features to reduce solar gain;
*   evaporative cooling systems and controls.

Table 6.4 Approaches to carbon reduction in existing non-domestic buildings in temperate climates

Building type	Peak internal heat gains (W/m^2)	Total energy consumption (MWh)		Carbon emission $(kgCO_2/m^2)$	
		Baseline	Post-intervention	Baseline	Post-intervention
Office: four-storey purpose built	Occupant – 5.4W/m² Lighting – 15.2W/m² Small power – 11.4W/m²	599	218	69	1 = 32 1 + 2 = 27 1 + 2+ 3 = 25
Office: five-storey converted warehouse	Occupant – 6.4W/m² Lighting – 16.9W/m² Small power – 11.2W/m²	413	149	60	1 = 27 1 + 2 = 27 1 + 2 + 3 = 24
Office: six-storey deep plan	Occupant – 5.4W/m² Lighting – 9.4W/m² Small power – 11.4W/m²	729	296	57	1 = 34 1 + 2 = 34 1 + 2 + 3 = 27
Office: small office in a terrace building	Occupant – 8.1W/m² Lighting – 15.8W/m² Small power – 25.7W/m²	34.6	15.8	128	1 + 2 = 66 1 + 2+3 = 63
Retail: small convenience store	Occupant – 15.0W/m² Lighting – 20.2W/m² Small power – 2.5W/m²	117	62	400	1 + 2 = 240 1 + 2 + 3 = 210
Retail: in a shopping centre	Occupant – 7.5W/m² Lighting – 19.8W/m² Small power – 3.3W/m²	80	35	93	1 = 58 1 + 2 = 51 1 + 2 + 3 = 41
Retail: large supermarket	Occupant – 9.6W/m² Lighting – 19.0W/m² Small power – 7.1W/m²	5,446	2,648	230	1 = 135 1 + 2 = 125 1 + 2 + 3 = 120
School: single-storey primary	Occupant – 11.3W/m² Lighting – 8.1W/m² Small power – 6.0W/m²	52	38	27	1 + 2 + 3 = 17
School: three-storey secondary	Occupant – 8.8W/m² Lighting – 8.5W/m² Small power – 4.9W/m²	547	358	25	1 = 16 1+2+3 = 14

Source: Based on data from Jenkins *et al.*, 2009.

Note: Intervention strategies in the last column: 1 = lighting and small equipment; 2 = fabric improvement; 3 = heating and controls. See discussions above for details on these three approaches.

Given the relatively low thermal mass of typical building envelopes in warm climates (especially in warm, humid ones) it is essential to 'condition' the outside air to minimise the temperature differences between the inside and outside. This will minimise the modulating role played by the building envelope, thus enabling the more energy efficient option of ventilation to cool buildings.

As pointed out in Chapter 5, buildings in low latitude (warm) climates are dominated by solar loading on the roof, as opposed to walls and floors, as is the case with buildings in high latitude (temperate/cold) climates. Therefore the external features needed to modulate the influence of a warm outdoors on the building indoor climate (and the attendant energy and carbon problems) are to reduce heat gain via the roof and reduce outdoor temperatures to facilitate smaller heat gain through the building envelope.

Even with all of the 'passive' approaches to climate-sensitive (and therefore low carbon) design, a warm climate poses significant challenges to thermal comfort serious enough to warrant some form of mechanical cooling. This is especially the case in existing buildings where the manoeuvrability of roof and external spaces for low carbon interventions is limited. Thus a sensible approach is to combine 'passive' design solutions with 'low energy' mechanical solutions to produce a passive *and* low energy architecture (PLEA).

Another approach is to take maximum advantage of the adaptive ability of humans in warm climates. A field study (Wijewardane and Jayasinghe, 2008) on thermal preference and thermal sensation in a light industrial environment in the warm, humid climate of Colombo, Sri Lanka found that free running (i.e. naturally ventilated) buildings could enhance their internal comfort by the following:

- Ensure the outdoor temperatures remain relatively low (extensive use of vegetation will be highly desirable; reservation of strips of sufficient width such as 10 metres or more between buildings for planting multiple rows of trees can be recommended).
- Use of insulation in the roofs and walls of buildings can be considered highly desirable. In order to ensure occupants are not subjected to radiation emitted by surrounding surfaces, the use of reflective surfaces with low emissivity facing the indoors will also be desirable.

In this detailed study, it was revealed that the factory workers involved in light work could feel reasonably comfortable even up to a temperature of 30.1°C when there is no significant air movement. This validated the use of a broader margin of about 3.5°C from the neutrality temperature for free running buildings accommodating people who are acclimatised to that particular climate. This could be further extended to take account of the physiological effect of cooling if sufficient air movement is available. When air velocity is high, the comfort zone can be extended by $6V - V^2$, where V is the indoor air velocity (Szokolay, 1991). For example, for an air velocity of 0.6 m/s, the upper temperature on the 50 per cent RH line can be increased by 3.24°C. The comfort surveys also validated the use of $6V - V^2$ as a good approximation to predict the temperature increase that is likely to be tolerated by people when higher air velocities are available. Both these are very important findings, since these guidelines can be used for establishing the comfort temperature for similar climates.

Manipulate external features

The low thermal mass typically found in warm climates together with the more open nature of buildings necessitates the conditioning of air outside the building as a key strategy to reduce energy demand to cool buildings in warm climates. Such 'climate proofing' of the outside could take the

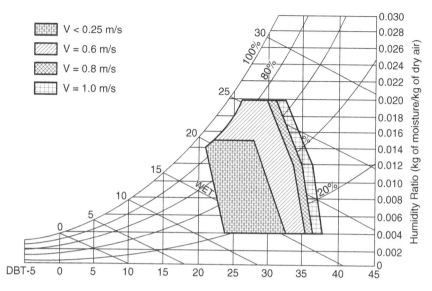

Figure 6.1 Extended comfort zone in warm, humid climates with ventilation

Source: Halwatura and Jayasinghe, 2008.

form of external shading, the use of vegetation, and the colour of external surfaces to reduce the climate burden on existing buildings. A detailed discussion of these strategies is given in Chapter 7.

Cooling systems and controls

Lowering of indoor temperatures by passive systems can be provided through the utilisation of several natural heat sinks: the ambient air, the upper atmosphere, water evaporation, and the under-surface cooled soil.

*   *nocturnal ventilative cooling:* lowering the indoor daytime temperature by ventilating the building at night;
*   *radiant cooling:* utilising nocturnal longwave radiant heat loss to the sky;
*   *direct evaporative cooling:* lowering the temperature while raising the humidity of the ventilation air by water evaporation;
*   *indirect evaporative cooling:* the primary cooling is derived from evaporation but the building is cooled indirectly, without elevating the indoor humidity.

Another approach is to cool a large volume of water evaporatively during the night hours and to use the cooled water during the day to cool the interior, usually by heat exchangers. The following approaches are possible:

*   a ventilated, shaded pond over an uninsulated concrete roof;
*   indirect evaporative cooling by roof ponds with floating insulation;
*   a roof pond;
*   a green roof;
*   outdoor ponds with floating insulation;
*   cooled soil as a cool layer.

6.2 Retrofitting buildings for low/zero carbon – climate-independent solutions

6.2.1 *Green roofs*

Green roofs act as a barrier to heat escaping from buildings (in cooler climates) as well as insulation against heat gain (in warmer climates) (see Table 6.5). They are therefore suitable for all climates, and their potential for use in building retrofits is beginning to receive wide attention (Castleton *et al.*, 2010). They are especially suited to older buildings that do not fulfil current best practices in building energy efficiency, provided the structural integrity of the roof to withstand the additional loads brought about by green roofs can be accommodated.

Stovin, Dunnett and Hallam (2007) state that the predominant constraint for a green roof is the load capacity of the existing roof structure. Buildings over 30 years old often have more reserve capacity than newer builds owing to the improved structural efficiency of modern analysis, design and construction methods. Stovin *et al.* (2007) retrofitted an extensive green roof on an office building in Sheffield to monitor *in situ* storm-water retention. The green roof was placed on two structural roof types: a reinforced concrete slab and profiled steel decking surfaced with plywood, without additional structural modifications in both cases. The concrete slab had an estimated capacity of $8–10kN/m^2$, enough to support a substrate depth up to 800 millimetres. This shows a strong case for retrofitting this building type in the UK.

Carter and Keeler (2008) calculated the net present value (NPV) of a green roof to be 10–14 per cent more expensive than a conventional roof over a 60-year lifetime. They identified that, if energy costs increase or green roof construction costs decrease or storm-water prevention becomes a higher public priority, then green roofs will become more economically attractive. They also noted that the positive social benefits of planting green roofs should not be overlooked and provide an additional incentive to the decision process.

Kosareo and Ries (2007) performed a comparative environmental lifecycle assessment of a green roof for a 1,115 square metre retail store in Pittsburgh, USA. They compared an intensive

Table 6.5 Heat partitioning in roofs

Heat loss pathways	Dry green roof (%)	Wet green roof (%)	Traditional roof (%)
Solar reflectivity	22.7	22.8	10.0
Solar absorption	38.6	38.6	
Outside adduction	23.7	12.9	85.6
Evapotranspiration	11.9	24.8	
Thermal accumulation	1.3	0.6	
Inside adduction	1.8	0.4	4.4

Source: Lazzarin, Castellotti and Busato, 2005.

Table 6.6 Energy savings of green roofs in warm conditions

Roof type	U-value (W/m^2K)		Annual energy saving		Total annual energy saving (%)
	Without green roof	With green roof	Heating (%)	Cooling (%)	
Well insulated	0.26–0.40	0.24–0.34	8–9	0	2
Moderately insulated	0.74–0.80	0.55–0.59	13	0–4	3–7
Uninsulated	7.76–18.18	1.73–1.99	45–46	22–45	31–44

Source: Niachou *et al.*, 2001.

green roof, an extensive green roof and a conventional ballasted roof. The increased roof lifetime of 45 years compared to the control roof lifetime of 15 years, along with the thermal conductivity of the growing medium, was found to have a significant impact on the lifecycle analysis. It was concluded that, although initial costs were high, the energy and cost savings made over the building lifetime meant that the green roof was an environmentally preferable choice. This is a more favourable outcome toward the green roof than that found by Carter and Keeler and highlights that the outcome of the lifecycle assessment of a green roof depends on the assumptions made for the calculations involved.

The special case of roofs in developing countries

In many developing countries corrugated metal roofs are very common. During the nights the low mass roof cools down rather quickly, acting in effect as an effective nocturnal radiator located directly above the living space. The indoor night conditions in such buildings are often more comfortable than in buildings with high mass, or with insulated, roofs. However, during the daytime hours the indoor climate in buildings with such roofs is often very uncomfortably hot, as the uninsulated metal roofs have much higher temperatures than a massive concrete roof. Installing under the roof centrally hinged insulating plates can reduce greatly the daytime heating without interfering too much with the cooling effect of such roofs during the nights. When the plates are in a horizontal position (closed), during the daytime, they form a continuous insulation layer under the roof, minimising the heat flow into the interior space. During the nights the plates should be turned into a vertical position, enabling radiant and convective heat flow from the interior space to the ceiling, which, in turn, is cooled by the radiation to the sky. Interior insulation plates are not exposed to the wind and the rain and thus can be simpler in construction, lighter, and much less expensive than external insulation panels. The changes in their position, vertical or horizontal, can be controlled from the interior manually, for example by a rope. A major potential hazard with interior insulation, if made of expanded plastic materials, is the risk of fire. A possible design of non-combustible operable interior insulation is wood frames with a lining of aluminium foil (Givoni, 2011).

6.2.2 Lighting and small appliances

Lighting and appliances represent a significant part of a building's energy consumption. In a typical UK office building, lighting and equipment account for 40 per cent of the total carbon emissions – office equipment 16 per cent, other electrical equipment 5 per cent and lighting 19 per cent (RAE, 2010). Lighting and appliances are therefore key to reducing the carbon footprint of existing buildings. The use of CFL bulbs and the switching off of unused lights and appliances are 'easy wins'. New technologies for energy efficient lighting are constantly emerging. Tubular fluorescent lighting is already exceeding 100 lm/W, and light emitting diodes (LED) are being championed as being the future of energy efficient lighting in all sectors (Jenkins *et al.*, 2009). This technology is predicted to exceed 150 lm/W by 2030 (Steigerwald *et al.*, 2002). Although lagging behind in terms of efficacy, organic light emitting diodes (OLED) are also showing potential, providing an even more versatile form of lighting that could be produced more cheaply and with lower embodied energy than conventional LED lights.

In terms of disadvantages, the colour rendering properties of current LEDs and OLEDs (i.e. achieving a 'white' light that is also energy efficient; Jenkins *et al.*, 2009) are questionable. Unless this property is improved, LEDs and OLEDs are not likely to achieve large scale acceptance in building applications.

Table 6.7 Likely cost and carbon savings from energy efficient lighting and appliances in the UK

Appliance	EU energy rating	Potential savings (£, per year)	Potential savings (kgCO₂/yr)
Fridge freezer	A+ or A++	£34	142
Upright/chest freezer	A+ or A++	£20	85
Refrigerator	A+ or A++	£12	48
Washing machine	A	£10	45
Dishwasher	A	£20	90
Integrated digital TV		£6	24
Energy efficient light bulb*		£4 (LED)	17
		£3 (CFL)	13

Source: EST, 2008a, 2008b; and Roaf *et al.*, 2008.

Notes: All figures are approximate and based on a gas centrally heated semi-detached house. All figures are for 2008.
* Based on replacing all light bulbs.

6.2.3 Smart meters

Smart meters are among the latest tools available to help reduce household energy consumption, and research has shown that they can save households between 3 and 15 per cent on their fuel bills. In the UK their current prices are around £40–£50; even a 5 per cent saving (equivalent to approximately £35) means that most meters can pay back their cost in about a year. As well as making households more aware of their energy use they also provide accurate data to energy companies. This reduces the need for estimate based billing and can be used to identify high energy consuming households and target them for energy efficiency schemes, as well as cutting operational costs. Making this data available to the government and to researchers (with the appropriate safeguards) would provide valuable evidence for research and policy making. The UK government is already committed to the installation of smart meters in all large business premises and is reviewing the case for domestic properties. Given the benefits of smart metering to households, energy suppliers, researchers and the government, it seems likely that this review will reach a positive conclusion.

6.2.4 CHP and renewable energy

Although it is not within the scope of this chapter to provide a detailed discussion of supply side options for reducing emissions in existing buildings it is worth noting that they will have a role to play in the domestic energy future. Micro-combined heat and power (CHP) schemes may be useful for dense groups of properties such as tenements and high rise flats, and community CHP schemes are already up and running in some UK councils (e.g. Aberdeen, Greenwich, Leicester, Southampton, Westminster and Woking; CHPA, 2011). Community CHP plants can be based at council premises, and therefore help reduce the emissions from both the council and the community it serves. Installing solar thermal and micro-generation systems such as photovoltaics and micro wind turbines is problematic for many individual properties owing to planning and conservation legislation. However, in light of the long term benefits of meeting energy demand and reducing emissions, future revisions of any housing legislation may lead to the removal of some of the barriers to stimulate the market and reduce the amount of disincentives for those householders considering investing in them.

6.3 Barriers and opportunities to LZC in existing buildings

Low carbon refurbishment of existing buildings faces many challenges, including legislative, environmental, cultural and financial drivers, as well as technical, financial and even legislative barriers. At the European level, two key pieces of legislation are influencing low carbon refurbishment. Firstly, there is the EU Renewable Directive with its binding target for renewable energy generation (in the case of the UK the target is 15 per cent of its energy consumption from renewable sources by 2020). Secondly, the Energy Performance of Buildings Directive (EPBD) seeks to realise prospective savings in the built environment, primarily through improving energy efficiency and developing a national calculation methodology for monitoring energy use. In May 2010 multiple amendments were made to the EPBD in order to increase the development and better understanding of LZC buildings (EC, 2010). In the case of the UK, additional legislative drivers include: the Low Carbon Transition Plan (DECC, 2009), which plots how the UK will achieve a 34 per cent carbon emission reduction by 2020 based upon the 1990 levels; and the Climate Change Act 2008, which is the legally binding framework that creates a new approach to managing and tackling climate change within the UK by setting an 80 per cent CO_2 reduction target by 2050.

The UK's case of retrofitting existing buildings to be low/zero carbon (or lack thereof) provides vital lessons for the drivers and barriers to LZC in existing buildings. The problem is widely recognised (as pointed out at the beginning of this chapter), yet action remains weak. Although the display of energy performance certificates (EPCs) has been mandatory throughout the UK since April 2006, there is currently no legislative driver to boost energy efficiency in the existing stock in the country.

6.3.1 Key drivers

A recent survey of architects engaged in refurbishment projects in England (Davies and Osmani, 2011) found that the key driver to LZC in existing buildings is 'cultural' (i.e. attitudinal or aspirational), bordering on belief rather than financial or legislative push factors. The following five statements elicited the highest affirmation:

1   Contribution towards sustainable communities for all stakeholders.
2   Refurbishment produces less waste and pollution for all stakeholders.
3   Refurbishment increases land conservation for all stakeholders.
4   Market strategy to secure low carbon refurbishment projects in the future for designers.
5   Refurbishment produces fast financial return for developers.

While it could be argued that item 5 above is 'financial' it is to be noted that the statement is true for refurbishment in general and not necessarily for LZC refurbishment. This is reflected in the UK government's own findings (DEFRA, 2005) which claimed that a behavioural change is much needed, as regulation and enforcement can progress only so far. Carter (2006) too echoed similar sentiments, stating that corporate social responsibility (CSR) policies have the potential to drive the production of LZC buildings.

However, it is to be expected that the effectiveness of behavioural change as a driver will be greatly enhanced by legislative, technical and financial incentives. Dobson (2007) recommended fiscal incentives; Lutzkendorf and Lorenz (2007) proposed the integration of sustainability factors in property valuations. Financial drivers, such as reduced operational energy costs and potentially higher rents and sales of refurbished buildings, could also be considered. Davies and Osmani's (2011) findings list the following 10 key incentives as relevant for LZC refurbishment in the UK:

1   a tax rebate for sustainable refurbishment projects;
2   a removed VAT difference between new build and refurbishment;
3   more research to produce affordable micro-generation technologies;
4   increased government supplied low carbon programmes and schemes;
5   increased government support of specific technologies and products;
6   simplified building regulations for future LZC refurbishment projects;
7   increased grants for large scale developers;
8   reduced cost differentiation between non-efficient and efficient products;
9   increased LZC refurbishment knowledge share and training for industry operatives;
10   development of a code for sustainable refurbishment.

The above incentives (drivers) could be grouped into three categories: financial (1, 2, 7 and 8); technical (3, 5 and 9); and legislative (4, 6 and 10).

6.3.2 *Key challenges*

Exploring the challenges facing low carbon renovation of existing houses in the UK, Lomas (2010) concluded that carbon reduction by way of energy efficiency improvement in the existing housing stock has the following key challenges:

1   Reducing energy use by implementing energy efficiency measures is more challenging than might be expected.
2   There is a shortage of information and tools by which the effectiveness of policy can be assessed.
3   Developing refurbishment strategies that target specific properties, such as larger detached properties, might improve cost-effectiveness.
4   Demand reduction initiatives might usefully address the design and the marketing of products and services.
5   Regulation may not be the appropriate mechanism for controlling energy use in the complex socio-technical system that is the occupied dwelling.

This was echoed by Davies and Osmani (2011), whose survey found financial and technical issues to be the top five challenges facing LZC refurbishment in the UK: high costs of micro-generation and energy efficient materials, and high taxation (financial challenges); and complexity of the existing building stock, and lack of trained site personnel (technical challenges). While the issue of LZC in new buildings, complex though it is, is well understood and legally and technically well backed up, refurbishment still has many technical and financial barriers. The case is likely to be more acute in developing countries, where information of stock conditions and trained personnel to undertake effective LZC refurbishment are likely to be even more scarce. These are further confounded by real data on energy and carbon performance in existing buildings. Real post-occupancy performance data and learning experience from many countries could improve our understanding of the scale of the problems faced by existing building stocks (Lomas, 2010).

References

All websites were last accessed on 30 November 2011.

Butler, D. 2008. Architects of a low-energy future. *Nature*, **452**, 3 April, pp. 520–523.
Carter, E. 2006. *Making Money from Sustainable Homes: A Developer's Guide.* Ascot: CIOB Publications.

Carter, T. and Keeler, A. 2008. Life-cycle cost–benefit analysis of extensive vegetated roof systems. *Journal of Environmental Management*, **87**, pp. 350–363.

Castleton, H.F., Stovin, V., Beck, S.B.M. and Davison, J.B. 2010. Green roofs, building energy savings and the potential for retrofit. *Energy and Buildings*, **42**, pp. 1582–1591.

Changeworks. 2008a. *Energy Heritage: A Guide to Improving Energy Efficiency in Traditional and Historic Homes*. Edinburgh: Changeworks. Available at: http://www.changeworks.org.uk/uploads/83096-EnergyHeritage_online1.pdf.

Changeworks. 2008b. Tenement fact sheet 2: draughtproofing of doors and windows, and between floorboards; secondary and double glazing. Available at: http://www.changeworks.org.uk/uploads/TFS_02.pdf.

CHPA (Combined Heat and Power Association). 2011. *CHP Users in the UK*. London: CHPA. Available at: http://www.chpa.co.uk/member-directory_42.html?SupplierCategory=1.

Climate Change Act 2008. Available at: http://www.legislation.gov.uk/ukpga/2008/27/contents.

Davies, P. and Osmani, M. 2011. Low carbon housing refurbishment challenges and incentives: architects' perspectives. *Building and Environment*, **46**, pp. 1691–1698.

DECC (Department of Energy and Climate Change). 2009. *The UK Low Carbon Transition Plan: National Strategy for Climate Change and Energy*. London: Stationery Office.

DEFRA (Department for Environment, Food and Rural Affairs). 2005. *Securing the Future: Delivering UK Sustainable Development Strategy*. Norwich: Stationery Office.

Dobson, A. 2007. Environmental citizenship: towards sustainable development. *Sustainable Development*, **15**, pp. 276–285.

EC (European Commission). 2010. *Energy Performance of Buildings Directive Amendments*. Brussels: EC. Available at: http://www.diag.org.uk/.

EST (Energy Saving Trust). 2008a. Energy saving light bulbs. Available at: http://www.energysavingtrust.org.uk/In-your-home/Lighting/Saving-energy-from-lighting.

EST. 2008b. *Energy Saving Assumptions*. London: EST. Available at: www.energysavingtrust.org.uk/content/download/30860/370167/version/1/file/EST+HEM+assumptions+doc.pdf.

EST. 2012. *Windows*. London: EST. Available at: http://www.energysavingtrust.org.uk/In-your-home/Roofs-floors-walls-and-windows/Windows.

Givoni, B. 2011. Indoor temperature reduction by passive cooling systems. *Solar Energy*, **85**, pp. 1692–1726.

Halwatura, R.U. and Jayasinghe, M.T.R. 2008. Thermal performance of insulated roof slabs in tropical climates. *Energy and Buildings*, **40**, pp. 1153–1160.

Jenkins, D., Banfill, P. and Pellegrini-Massini, G. 2009. *Non-Domestic Conclusions of the Tarbase Project: Reducing CO_2 Emissions of Existing Buildings*. Edinburgh: Urban Energy Research Group, School of Built Environment, Heriot-Watt University. Available at: http://www.sbe.hw.ac.uk/documents/TARBASE_ND_REPORT.pdf.

Kosareo, L. and Ries, R. 2007. Comparative environmental life cycle assessment of green roofs. *Building and Environment*, **42**, pp. 2606–2613.

Lazzarin, R.M., Castellotti, F. and Busato, F. 2005. Experimental measurements and numerical modelling of a green roof. *Energy and Buildings*, **37**, pp. 1260–1267.

Lomas, K. 2010. Carbon reduction in existing buildings: a trans-disciplinary approach. *Building Research and Information*, **38**, pp. 1–11.

Lutzkendorf, T. and Lorenz, D. 2007. Integrating sustainability into property risk assessments for market transformation. *Building Research and Information*, **35**, pp. 644–661.

Niachou, A., Papakonstantinou, K., Santamouris, M., Tsangrassoulis, A. and Mihalakakou, G. 2001. Analysis of the green roof thermal properties and investigation of its energy performance. *Energy and Buildings*, **33**, pp. 719–729.

RAE (Royal Academy of Engineering). 2010. *Engineering a Low Carbon Built Environment*. London: RAE. Available at: www.raeng.org.uk.

Roaf, S., Baker, K. and Peacock, A. 2008. *Evidence on Hard to Treat Properties*. Edinburgh: Scottish Government. Available at: www.scotland.gov.uk/Publications/2008/10/17095821/0.

Steigerwald, D.A., Bhat, J.C., Collins, D., Fletcher, R.M., Holcomb, M.O., Ludowise, M.J., Martin, P.S. and Rudaz, S.L. 2002. Illumination with solid state lighting technology. *IEEE Journal on Selected Topics in Quantum Electronics*, **8**, pp. 310–320.

Stovin, V., Dunnett, N. and Hallam, A. 2007. Green roofs: getting sustainable drainage off the ground. In *6th International Conference of Sustainable Techniques and Strategies in Urban Water Management (Novatech 2007), Lyon, France, 2007*, pp. 11–18.

Szokolay, S.V. 1991. Heating and cooling of buildings. In *Handbook of Architectural Technology*, ed. H.J. Cowen. New York: Van Nostrand Reinhold.

Walker, S., Cairns, P., Cormack, D., Máté, I., McLaren, D. and Futak-Campbell, B. 2010. *Scottish House Condition Survey 2009*. Edinburgh: Scottish Government. Available at: http://www.shcs.gov.uk.

Wijewardane, S. and Jayasinghe, M.T.R. 2008. Thermal comfort temperature range for factory workers in warm humid tropical climates. *Renewable Energy*, **33**, pp. 2057–2063.

Wilkinson, P., Smith, K.R., Beevers, S., Tonne, C. and Oreszczyn, T. 2007. Energy, energy efficiency, and the built environment. *Lancet*, **370**, pp. 1175–1187.

7   Carbon management in cities

7.1   Introduction

Whichever way we count the emission of GHGs (i.e. production based or consumption based), urban areas account for the lion's share of the world total emissions. A production based apportionment (i.e. allocating emissions to those who generate them) of GHGs will lead to approximately 71 per cent of the total emission being attributed to urban areas (IEA, 2010). If a consumption based apportionment (where emissions are allocated to those whose consumption caused the emissions) is used, urban share of the GHG emissions will be much higher, since emissions from a whole range of sectors (such as agriculture, forestry and commodities) will have to be included (Hoornweg, Sugar and Gómez, 2011). While it may be tempting to assign 'emission blame' to cities, it must be remembered that cities drive the national economies of most nations. In other words, urban consumption (and therefore associated emissions) benefits not only cities but also the countries where cities are located (see Dodman, 2009).

Owing to the complexities involved in obtaining fine grained data at local level, the analysis of carbon in cities and actions to mitigate it is relatively new. Most studies focus on single cities or cities in a single country. What seems to emerge is that the divergent carbon profiles of cities more or less coalesce into two broad categories:

1   In more developed economies, large cities are generally more carbon efficient than smaller towns or rural areas. For example, Brown, Southworth and Sarzynski (2008) found that the average metropolitan area resident in the US had a partial carbon footprint (from residential energy use plus transportation) of 2.24 metric tons in 2005, which was 86 per cent of the average American's partial footprint (2.60 metric tons). This was largely due to less car travel and residential electricity use, while rural areas needed more freight travel and residential fuels.

2   Large urban centres in the developing world have higher emissions than smaller cities or rural areas (owing to an increased standard of living and purchasing power, greater use of private vehicles, rapid urban sprawl and more protein rich and energy intensive diets; see Lebel *et al.*, 2007).

At the same time, the climatic contexts of cities, the presence of stronger or weaker urban heat islands, the transport networks and modal splits, and the extent and proximity of the rural hinterland complicate the urban carbon footprint greatly. Cities in temperate climates are dominated by heating energy need, which can be efficiently supplied when buildings are clustered together; in arid and tropical climates, cooling dominates energy need, and cooling is inherently more inefficient.

7.1.1 *Data on global cities and their carbon emissions*

The GHG emissions of a city are strongly dependent upon its location and the economic situation of the country in which the city is located. Climate, in particular heating degree days, is currently an important determinant of the amount of energy required to heat urban buildings. This is likely to change with increasing affluence in warm climate cities, where greater use of air conditioners (likely to increase with global warming) could make cooling degree days equally, if not more, important. This is evident from the USA, where household electricity use rises sharply with average July temperature (Glaeser and Kahn, 2010). Moreover, the location of a city often determines its status as a gateway, thereby explaining emissions arising from aeroplanes and shipping (Kennedy *et al.*, 2009).

Analysing the carbon footprints of 12 global cities, Sovacool and Brown (2010) concluded that four factors account for the majority of differences in carbon footprint between cities:

1   income (especially per capita purchasing power);
2   population density and compactness (especially population and employment densities, and mixed land use) – a more accurate urban planning term that is more closely relevant to carbon management is 'urban form' (Lebel *et al.*, 2007);

Table 7.1 Urban GHG emissions from selected global cities

City	Country	Urban per capita emission (tons)	Country per capita emission (tons)	Urban population* (millions)
Chitagong	Bangladesh	0.1	0.4	2.9
Kathmandu	Nepal	0.1	1.5	0.7
Thimpu	Bhutan	0.3	2.5	0.1
Dhaka	Bangladesh	0.6	0.4	7.7
Kolkata	India	1.1	1.3	4.6
São Paulo	Brazil	1.4	4.2	11.1
Delhi	India	1.5	1.3	9.9
Colombo	Sri Lanka	1.5	1.6	0.7
Rio de Janeiro	Brazil	2.1	4.2	6.3
Mexico City	Mexico	2.8	5.5	8.6
Amman	Jordan	3.2	4.0	1.0
Stockholm	Sweden	3.6	7.1	0.8
Buenos Aires	Argentina	3.8	7.6	3.1
Seoul	South Korea	4.1	11.5	9.8
Barcelona	Spain	4.2	9.9	1.6
Vancouver	Canada	4.9	22.6	2.1
Tokyo	Japan	4.9	10.8	8.9
Paris	France	5.2	8.7	2.2
Madrid	Spain	6.9	9.9	3.3
Helsinki	Finland	7.0	14.8	1.1
Brussels	Belgium	7.5	12.4	0.2
Cape Town	South Africa	7.6	9.9	2.4
Glasgow	UK	8.8	10.5	0.6
Toronto	Canada	9.5	22.6	5.1
London	UK	9.6	10.5	7.6
Hamburg	Germany	9.7	11.6	1.8
Turin	Italy	9.7	9.3	0.9
Beijing	China	10.1	3.4	10.3
Athens	Greece	10.4	11.8	3.1
New York	USA	10.5	23.6	8.2
Bangkok	Thailand	10.7	3.8	6.4
Shanghai	China	11.7	3.4	14.2
Portland	USA	12.4	23.6	0.6
Los Angeles	USA	13.0	23.6	3.8
Seattle	USA	13.7	23.6	0.6
Austin	USA	15.6	23.6	0.8
Stuttgart	Germany	16.0	11.6	0.6
Sydney	Australia	20.3	25.8	4.6
Denver	USA	21.5	23.6	0.6
Rotterdam	Netherlands	29.8	12.7	0.8

Source: Per capita emission data based on Hoornweg, Sugar and Gómez, 2011; urban population data from UNPD, 2010 and Brinkhoff, 2011.

Note: * 'Urban population' in the last column refers to the city population within municipal boundaries, not in the agglomeration.

3 modes of transport (especially efficient mass transit and promotion of walking and cycling);
4 electricity supply (especially reliance on renewable technologies; see Table 7.2);
5 trade-offs, rebounds and other effects (especially policy priorities targeting the wrong sectors, interaction between the above four factors, etc.).

Income

Income levels explain a large amount of variations in urban carbon footprint. As can be seen in Table 7.1, poorer cities such as Dhaka and Kolkata have some of the smallest carbon footprints, and wealthier cities such as Sydney, Rotterdam and Denver have over 20 times larger carbon footprints than their poorer counterparts. However, the relative performance of cities compared to their national average carbon footprint appears to be equally important. In poorer countries (such as India and Bangladesh) cities have a larger carbon footprint than their national average (Lebel *et al.*, 2007), whereas several cities in developed economies have a smaller carbon footprint than their national average (Glaeser and Kahn, 2010). In fact, the emission profiles of more established cities in the US are strikingly different to those of the suburbs (Glaeser and Kahn, 2010). The case of relatively wealthy cities in poor countries may be better explained by income, while the latter (developed cities in wealthier countries) have urban morphological, transport and other causes for their superior carbon performance.

Urban form

Sprawling urban growth, in which large areas of land are hard to access for either residential or agricultural use, makes personal vehicles essential. The increasing use of personal vehicles leads to greater demand for roads, which is quickly filled up by even greater use of personal vehicles. As a result, average speed for all vehicles drops, idling times increase and fuel consumption for travel (and associated pollution) increases.

The key variables of urban form that are relevant to the urban carbon footprint are density (population, housing and jobs), compactness, and road layout and connectivity (see Lebel *et al.*, 2007; Sovacool and Brown, 2010).

At higher population densities, human interactions needed to sustain life can occur with little carbon emission (Jabareen, 2006). Higher housing density leads to fewer vehicle miles travelled; furthermore people living in high housing density areas tend to buy more fuel efficient (i.e. smaller) vehicles (Golob and Brownstone, 2005). Higher job densities make fuel efficient transport more sustainable, thus leading to lower emissions.

Compactness implies multiple uses for a given parcel of land. Compactness mitigates urban sprawl, promotes intensification and density of development and activity, and prevents rural areas from becoming suburban satellites. Compactness also increases social interaction and access to available energy services, and reduces energy consumption by providing buildings more amenable to district heating and combined heat and power. Compact cities reduce the number and length of transport trips, and are more suitable for bicycling and walking (Jabareen, 2006).

Road layout and connectivity determine the mode, intensity and distance of travel, which contribute to greater or lesser carbon emission. Cities that are concentric (i.e. those with a strong core with radial arteries emanating from the core) tend to have inordinate congestion on their arterial routes, leading to greater fuel waste.

Modes of transport

Cities that rely on private automobiles for a majority of the travel tend to have the highest carbon emission. Modes of transport are closely linked to 'urban form'. Cities that promote walking, cycling and efficient public transportation tend to have a smaller carbon footprint associated with the transportation sector.

Apart from the interactions between modes of transport and urban form, carbon emission due to transportation is also influenced by income. The developing world, especially in Asia, is currently witnessing unparalleled growth in private automobiles, leading to a larger carbon footprint. This partly explains the larger urban carbon footprint in developing cities relative to their national average.

Electricity supply

A more significant determinant of GHGs from electricity is the means of power generation as well as the nature and severity of the climatic zone where the city is located. Access to hydropower substantially reduces the intensity of emissions from these cities. Cities located close to abundant coal seams (for example, Prague; Kennedy *et al.*, 2009) have some of the highest emissions (see also Table 7.2).

Table 7.1 and the above discussions indicate that, in general, low and middle income countries tend to have lower per capita emissions than high income countries; dense cities tend to have relatively lower per capita emissions (particularly those with good transportation systems); and cities in colder climates tend to have higher emissions. The most important observation is that there is no single factor that can explain variations in per capita emissions across cities; they are agglomerations of a variety of physical, economic and social factors specific to their unique urban life (Hoornweg *et al.*, 2011). However, the level of development may be a critical factor; thus our subsequent discussions look at strategies to reduce urban GHG emissions by the developmental contexts of cities.

7.2 Broad strategies for urban carbon management

Broadly speaking, carbon emission in cities comes from three sources:

- buildings;
- transport;
- waste.

In a few cases agriculture may play a small but significant part, but this is not the case across all cities. Table 7.3 shows the sectoral decomposition of carbon emission in 12 major global cities.

It is clear that buildings dominate the carbon emission profiles in most cities. Where they appear low, the reason tends to be excessive energy use for transport (as in Manila, Delhi and São Paulo). In terms of building emissions, electricity generation is the key driver, while private automobiles are the principal driver for urban transportation emissions. While usually small, emission from waste is dominated by landfills. Thus electricity generation, private automobiles and landfills ought to be the principle focus areas for urban carbon management. Demand reduction in terms of building energy use, private automobile travel and landfills will greatly reduce the urban carbon footprint.

Reducing the carbon emission from energy generation is extensively covered in Chapter 4. Energy demand reduction in buildings is extensively covered in Chapters 5 (new build) and 6

Table 7.2 Lifecycle equivalent carbon emission from electricity generation

Source of electricity	Lifecycle equivalent carbon emission (g of CO_2 per kWh)
Renewable sources (falling water, wind, solar and geothermal)	5.1–59.6
Nuclear	124
Clean coal with carbon capture and storage	439
Conventional fossil fuel	443

Source: Based on data from Jacobson, 2009.

Table 7.3 Sectoral decomposition of carbon emission in major global cities

Metropolitan area	Sectoral carbon emission			
	Buildings and industry (%)	Transport (%)	Agriculture (%)	Waste (%)
Beijing	87*	5	1	1
Jakarta	56	41	<1	<1
London	76	23	<1	<1
Los Angeles	52	48	<1	<1
Manila	39	51	9	1
Mexico City	45*	35	6	<1
Delhi	32	66	2	<1
New York	77	23	<1	<1
São Paulo	24	51	2	23
Seoul	44*	42	1	13
Singapore	83	17	–	–
Tokyo	67	32	<1	<1

Source: Sovacool and Brown, 2010.

Notes: * Figures in the first column for these cities refer mainly to industrial emissions; buildings dominate in all other cities. Total may not always be 100% owing to missing data and/or rounding errors.

(refurbishment). What remains to be discussed is the *urban* strategies to reduce built environment emissions. In this regard, a key approach is to mitigate the local warming phenomenon known as the urban heat island. Efficient waste management is the other key urban strategy for a low carbon built environment.

Urban heat island and carbon emission

An urban heat island (UHI) is best visualised as a dome of stagnant warm air, over the heavily built-up areas of cities. UHIs have been observed in practically all parts of the world except in extremely cold climates. UHIs are especially important in warm cities or cities with warm summers. UHIs are intense at night, occurring a few hours after the sunset.

A clear indication of a UHI is the rapid rise in minimum (night-time) temperature and a slow rise in maximum (daytime) temperature: thus the daily (i.e. diurnal) variation between the maximum and the minimum temperatures will diminish over time (Landsberg, 1981: 87).

Table 7.4 Characteristics of an urban heat island

Climatic parameter	Effect of urbanisation
Temperature	Rise in daily minimum temperature: some change in maximum temperature.
Humidity	Reduction in daytime humidity, but increase in night-time values.
Precipitation	Larger increases in summer (up to 21%) and smaller increases in winter (5–8%). In the tropics, the increase is attributed more to air pollution than heat emission.
Wind	Increases in the number of calm periods observed. Up to 20% reductions in wind speeds are known. The effect is greater upon weaker winds.
Solar radiation	Though incoming radiation values are not changed, the apparent values are high owing to the containment of reflected radiation by the heat dome.

Source: Emmanuel, 2005.

The causes for UHIs range from urban geometry to work week patterns, from anthropogenic heat to thermal characteristics of urban surfaces, and from obstruction to wind flow to lack of vegetation. However, the widely prevalent view among urban climatologists is that, at neighbourhood and smaller scales, urban geometry leads the list of possible causes for the heat island phenomenon (Emmanuel, 2005).

7.3 Carbon management in developed cities

In developed cities, buildings account for the largest amount of GHG emissions (up to 85 per cent of all emissions; see Table 7.4). Reducing the emissions in the built environment, using the strategies outlined in Chapters 5 and 6, therefore is the top priority for developed cities (while it forms an important part of the policy options in developing cities as well). The second largest urban emissions come from transportation – and these can be reduced only by planning the city morphology at a climate friendly level. Lifestyle factors such as waste generation are the third important platform for GHG reduction in developed cities.

7.3.1 Shelter

While the main strategies are listed in Chapter 5 and 6, this section outlines the 'urban' shelter related strategies that could be employed to reduce building related GHG emission from cities.

A key to carbon management of the built environment in developed cities is to recognise that many of these cities are shrinking in population. 'Shrinking cities' – a concept initially theorised in the wake of German unification (Rieniets, 2009) – is an increasingly common reality in many parts of the world. Over the last 50 years, 370 cities throughout the world with populations over 100,000 have shrunk by at least 10 per cent (Oswalt and Rieniets, 2007). These are more common in the industrial heartlands of the USA (59 cities), Britain (27), Germany (26), Italy (23), Russia (13), South Africa (17) and Japan (12). They are also common in other parts of the world, even as growing cities continue to dominate the discourse. A typical planning approach to this crisis is to reconceptualise decline as shrinkage and to explore creative and innovative ways for cities to shrink successfully. Such approaches have usually taken the form of land for recreation, agriculture, green infrastructure and other non-traditional land uses beneficial to existing residents and that will attract future development (Hollander *et al.*, 2009).

In their drive towards being sustainable and ecologically sound places, shrinking cities will need to consider the local climate implications of their current urban trajectories. While population may decline, the underlying urban morphology largely remains in place, leading to the continuation of the urban climate anomaly. However, in the case of cool climate cities such as Glasgow this aspect of shrinking is beneficial (Emmanuel and Krüger, 2012). Urban warmth created by a judicious arrangement of land use and land cover (as evidenced by the appropriate local climate zone class) could be exploited for energy efficient uses such as district heating and to enhance the feasibility of low carbon options such as district ground source heating or other communal renewable technologies. Emmanuel and Krüger (2012) show that the UHI itself does not go away, even in shrinking cities; thereby the opportunities to be sustainable and low carbon might still be available. At the same time, the summertime trends suggest that overheating may become a distinct possibility in the future. These realities should inform shrinking cities in their attempt to re-invent themselves in a carbon and energy efficient fashion.

7.3.2 *Mobility*

Cities that promote walking and/or cycling, as well as public transit systems such as mass transit systems (MRTs), tend to leave their residents with less need to purchase private automobiles. This leads to efficient energy use for mobility. (See Table 7.5 for the carrying capacity for different modes of transport.) Furthermore, the promotion of walking and cycling reduces the need for on-site parking, which leads to higher density (which in turn favours public transit). Walking and cycling usually tend to promote neighbourhoods with more pedestrian friendly features and a more connected street layout (Holtzclaw, 2004).

7.3.3 *Lifestyle*

A simple provision of mass transit systems, effective tackling of climate based building energy needs by manipulating the urban form, and other planning strategies do not guarantee an automatic reduction in urban GHG emissions. The energy use for (and therefore emissions from) transportation and waste in many developed cities with 'enlightened' public policies is a case in point. For example, the highly developed city state of Singapore makes it very difficult to own and maintain a car. In addition to the 'market value' of a car (called the open market value, or OMV, consisting of manufacturer's price, freight, insurance, overheads and so on) the following are also payable:

- ARF (additional registration fee): 110 per cent of OMV;
- CoE (certificate of entitlement): all vehicle owners must possess a CoE, the price for which is determined by an annual auction system;

Table 7.5 Energy use by passenger vehicles in the USA

Mode	Load factor (passenger km/vehicle km)	Relative energy use (intercity bus = 1)
Intercity bus	41.8	1.00
Intercity rail	19.1	2.12
Commuter rail	35.6	2.19
Transit bus	12.7	2.43
Transit rail	23.1	2.77
Automobile	1.7	3.20
Air: certified route	89.3	3.82

Source: Emmanuel, 2005.

Table 7.6 Transport characteristics in four archetypical US neighbourhoods

	Sprawl suburb	Commuter village	City centre	Major metropolitan centre
Residential density (households/residential Ha)	7.9	24.7	247	494
Transit (no. of buses/hr)	1	27	90	Very high
Shopping (% of houses that have at least five shops within walking distance)	None	25%	100%	100%
Pedestrian amenities	Low	Medium	High	High
Automobiles per capita	0.79	0.66	0.28	0.12
Automobile kilometres travelled	16,945	10,328	4,414	1,832
Annual household automobile cost	$8,200	$5,030	$1,900	$800

Source: Based on data from Holtzclaw, 2004.

- excise duty: 45 per cent of OMV;
- GST (goods and services tax): 5 per cent of cost including excise duty;
- registration tax: $140;
- road tax;
- transfer fee: 2 per cent of the vehicle's value.

The above costs could make a standard car's final price rise by seven or eight times the 'market value' of the car. Money raised by the system is invested in the public transit system, which is excellent by any standards. Nevertheless, the rate of growth in automobiles in Singapore has exceeded 3 per cent per annum since the introduction of the vehicle quota system in Singapore in 1990 (although the growth has been halved in the last three years owing to fewer vehicle quotas being released by the Government of Singapore; see Singapore Government, 2011). A carrot and stick policy utilising tax regime and public transport policy has not led to a substantial reduction in automobile ownership in Singapore (in fact, Singapore has one of the highest vehicle densities among cities – more than 210 motor vehicles per kilometre of road, compared to London at 70 vehicles per kilometre of road, Tokyo at 45, and the United States at 33). The reasons are many: the car as a social symbol (especially when it is extremely expensive to own one); privacy; the perceived discomfort of public transport in a hot and humid environment, and so on. Similarly, lifestyle choices explain the amount of waste generated (and waste recycled) in developed economies.

Case study: London

The London Mayor's Climate Change Action Plan (Mayor of London, 2007) provides one of the earliest city-wide urban carbon management action plans. London is responsible for 8 per cent of the UK's emissions, producing 44 million tonnes of CO_2 each year. Given its economic position and population growth, London's emissions are projected to increase by 15 per cent to 51 million tonnes by 2025.

London's plan to tackle its emission starts from the broad view that this should be achieved through a process of 'contraction and convergence' – with the largest industrialised nations that have caused climate change required to reduce their emissions significantly, while newly developing nations are permitted to increase emissions up to a point where emissions converge and stabilise at a level which avoids catastrophic climate change. The plan assumes that stabilising atmospheric CO_2 concentrations at 450 parts per million (ppm) is required to avoid catastrophic climate change. Stabilising global carbon emissions at 450ppm on a contraction and convergence basis means that London has to limit the total amount of carbon dioxide it produces between now and 2025 to about 600 million tonnes. Meeting this CO_2 budget will require on-going reductions of 4 per cent per annum. This implies a target of stabilising London and the UK's emissions at 60 per cent below 1990 levels by 2025.

The achievement of the above target is expected to be achieved by the following actions.

Emissions from existing homes

Energy use in existing homes is the largest single source of CO_2 emissions in London, at nearly 40 per cent of the total. Approximately 7.7 million tonnes needs to be removed by 2025.

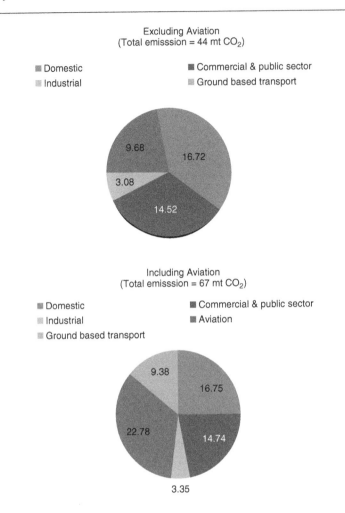

Figure 7.1   London's CO_2 emissions in 2006

Source: Based on data from Mayor of London, 2007.

The Mayor's Green Homes Programme includes:

- A London-wide offer to homeowners of heavily subsidised (and free to those on benefits) loft and cavity wall insulation.
- A major marketing campaign to increase awareness about what actions Londoners can take to cut their emissions and reduce their energy bills.
- A new one-stop-shop advice and referral service, available to all Londoners, on implementing energy savings measures and installing micro renewables, which will be accessible by web and by phone. This pioneering service will be delivered in partnership with the Energy Savings Trust.
- A pilot Green Homes 'concierge service', providing bespoke energy audits and project management of installation of energy efficiency improvements, micro renewables and water conservation measures for the able-to-pay sector.

- A programme of improving the energy efficiency of London's social housing stock.
- Identifying skills gaps in the sustainable energy industry and developing training (in collaboration with the relevant industry bodies) to improve the skills required to install and service energy saving and micro renewable products and systems. Implementing the Mayor's home energy efficiency programme would save the average London household £300 per year.

Emissions from existing commercial and public sector activity

Emissions from the commercial and public sector are 15 million tonnes of CO_2 annually (18 million when industry is included). They come primarily from electricity usage, including for lighting and computing, although, as the climate continues to warm, energy used for cooling buildings could become increasingly significant. Substantial savings can be achieved through simple actions like turning off appliances at night and avoiding inefficient heating and cooling of buildings. These carbon savings will also result in significantly lower energy bills, and will boost London's economy and create new jobs by creating demand for services such as energy saving building refurbishment.

The Mayor's Green Organisation Programme includes:

- *The Better Buildings Partnership:* working with and incentivising commercial landlords to upgrade their buildings, particularly during routine refurbishments.
- *The Green Organisations Badging Scheme:* working with tenants (both private and public sector organisations) to reduce emissions through staff behavioural changes and improved building operations. This will include providing information and support to deliver these changes, working together with existing initiatives, as well as a clear set of targets and associated green 'badging' levels.
- *Lobbying:* Both the Better Buildings Partnership and the Green Organisations Badging Scheme will be supported by a lobbying campaign focusing on key barriers to the uptake of energy savings and clean energy. The Mayor will work closely with London's businesses and public sector organisations to develop and deliver this programme, in order to build on and benefit from initiatives that are already in train.

Emissions from new build and development

Roughly 1 million tonnes of CO_2 per annum can be saved by 2025 through better enforcement of current regulations and the introduction of higher standards for domestic and commercial new build. The major challenge will be to ensure concerted action by all organisations involved (including boroughs, developers and the construction industry) and full implementation of the improved standards.

- *To revise the London Plan requirements for new developments:* The draft Further Alterations to the London Plan issued by the Mayor require new developments to prioritise the use of decentralised energy supply, most importantly by connecting to combined cooling, heat and power (CCHP) networks.
- *Further emphasis on energy efficiency through the Mayor's planning role:* Recent experience within the Greater London Authority has demonstrated what a substantial difference

a small number of additional dedicated and knowledgeable resources can make. New energy focused resources will therefore be added to both the Mayor's Planning Decisions Unit and the Environment Team.

- *A greater focus on energy efficiency at the borough level:* to increase sustainable energy and planning skills in the London boroughs and among other key stakeholders through a comprehensive outreach programme. This programme will provide training and support for the boroughs, a publicly accessible energy portal and close collaboration with developers to establish the true cost–benefit of compliance with new regulations in London.
- *Showing by doing:* individual developments and new housing powers. The Mayor will model exemplary energy efficiency standards both through individual developments in which the London Development Agency (LDA) is involved, and for all new affordable homes. The Mayor's new housing strategy will make energy efficiency a key priority, including achieving the UK government's target of 100 per cent new homes as zero carbon by 2016. All LDA developments will also be developed to the highest standards, building upon experiences such as the Gallions Park zero carbon development.

Emissions from the energy supply

The single biggest barrier to reducing London's carbon emissions is the way in which energy supplied to homes and offices is produced and distributed. Centralised electricity generation, whether through coal, oil, gas or nuclear power stations, is inherently inefficient – wasting two-thirds or more of its original energy input in the form of expelled heat. Further losses occur in the process of distributing electricity from rural power stations to the towns and cities where it is mostly consumed.

The Mayor's top priority for reducing carbon emissions is to move as much of London as possible away from reliance on the national grid and on to local, lower carbon energy supply (decentralised energy, including CCHP, energy from waste, and on-site renewable energy, such as solar panels). This approach is often termed 'decentralised energy'.

- Dramatically increasing the roll-out of a CCHP energy supply. The main source of carbon reductions from decentralised energy will come from the combined generation of heat and power locally. Through the direct investment of the London Development Agency and the requirements of the draft Further Alterations to the Mayor's London Plan, supplying energy through CCHP will become the norm in major new developments in London. However, the bulk of CCHP's potential will need to be realised through supplying London's existing building stock. A major vehicle for this will be the Mayor's Climate Change Agency and its joint venture with EDF Energy, the London Energy Services Company.
- Rapidly developing and delivering mechanisms to produce energy from waste (without incineration). Energy from waste through new non-incineration technologies (such as anaerobic digestion, mechanical biological treatment, pyrolysis and gasification) offers a carbon savings potential nearly as large as that of CCHP. If all the London waste that currently goes to landfill were utilised, it could generate enough electricity for up to 2 million homes, and heat for up to 625,000 homes. However, this technology still needs to be explored and commercialised. The creation of a single

waste authority for London would have provided a major boost to this work. In its absence, the Mayor will work with boroughs and industry to facilitate and accelerate the potential of energy from waste, including pilots and showcasing technologies at several large sites.

- Promoting the uptake of on-site renewable energy in London. Small and medium scale renewable energy generation will be promoted through the revised London Plan standards, the Green Homes and Green Organisations Programmes, and the mayoral group's own installations.
- Pursuing large scale renewable power generation in London. There are limited, but significant, opportunities for large scale renewable power generation in London; for example, land based wind turbines could supply power to up to 47,000 households. Much greater opportunities for wind power exist in the Thames Estuary, at least enough to supply a million homes, and the Mayor will strongly back projects such as the London Array. London will also investigate the potential for using tidal and wave power from the Thames.
- Making the case for a greatly accelerated programme of investment in renewable energy in the UK. The UK has huge untapped potential for renewable energy. In fact, some estimates suggest that renewables could provide nearly 100 per cent of the UK's electricity, with off-shore wind providing up to 60 per cent of the total. This would be one of the measures with the most significant carbon reducing impact for London. London will therefore push for simple planning and regulatory changes that incentivise a much greater contribution from renewables to the national grid.
- Supporting carbon sequestration. While burying carbon emissions underground is not a long term solution, the Mayor recognises that it is being investigated around the world and will offer significant emissions reductions once commercially viable. London will push for rapid uptake of carbon sequestration to reduce national grid emissions further while the UK achieves a transition to a renewable energy based economy. In the meantime, all new power stations should use the latest technologies to minimise CO_2 emissions and implement heat capture and distribution.

Emissions from ground based transport

London is unusual compared with many large cities around the world in that its emissions from transport (excluding aviation) are relatively small – about 23 per cent of the total. Unlike other sectors, transport emissions in London have stayed flat since 1990 despite the rapid growth of London's population and economy. This is thanks to high long term levels of public transport use and, since 2000, unprecedented investment in the public transport network, alongside the implementation of policies like the congestion charge to combat congestion and manage traffic.

If implemented, the measures in this plan would deliver carbon savings of 4.3 million tonnes by 2025.

The top priority now is to reduce emissions from car and freight traffic, since these represent nearly three-quarters of emissions in this sector. This includes changing the way Londoners travel, and a major programme of continued investment in public transport, walking and cycling to provide attractive alternatives to car travel (as outlined in Transport for London's Transport 2025 work). It also includes promoting alternatives to the car through marketing, information and other travel demand management policies. London-wide, this can deliver:

- *A million tonnes of CO$_2$ savings per annum:* For an average Londoner, switching from driving to work to taking the bus will save 0.6 tonnes of carbon per year; taking up cycling instead would increase these savings to 1.1 tonnes.
- *Operating vehicles more efficiently:* Simply driving more sensibly can reduce fuel use by 5–10 per cent. The Mayor will promote eco-driving (for example, smoother acceleration and braking and proper vehicle maintenance) by all car, freight, taxi and public transport drivers.
- *Promoting low carbon vehicles and fuels:* The biggest opportunity for emissions reductions in this sector is from the uptake of lower carbon vehicles and fuels, which alone could cut transport emissions by up to 4–5 million tonnes. CO$_2$ emissions from road transport would fall by as much as 30 per cent if people simply bought the most fuel efficient car in each class.
- *Carbon pricing for transport:* More widespread carbon pricing will be essential to incentivise demand for low carbon vehicles and fuels, and to drive innovation in further developing these technologies. Comprehensive carbon pricing requires regulatory changes at international, national and regional levels. Having led the world with the Central London congestion charge, the Mayor now wants London to become the first major city in the world to charge cars to enter its central business area on the basis of their carbon emission levels. Under this proposal, the highest polluting vehicles will be charged £25 a day, while zero emission vehicles will travel free. The Mayor will also pursue an ambitious programme of energy saving measures across public transport. This includes regenerative braking on the Tube, which allows energy generated in braking to be reused to drive the train, and the conversion of London's entire 8,000-bus fleet to diesel electric hybrid vehicles.

Table 7.7 Summary of London's approach to a low carbon future

Sector	Identified saving by 2025	Strategic focus	Policy approach to 2025 target	Contribution to identified saving (%)
Domestic	39% (7.644 mtCO$_2$)	Emission from existing homes	Lighting and appliance	23
		Emission from energy supply	Energy supply	44
			Behaviour change	18
			Thermal efficiency	10
		Emission from new build and development	New build	5
Commercial and public sector	39% (7.644 mtCO$_2$)	Emission from existing commercial and public sector	Energy supply	50
		Emission from energy supply	Behaviour change	25
			Improved physical infrastructure	20
		Emission from new build and development	New build	5

Ground based transport	22% (4.312 mtCO$_2$)	Emission from ground based transport	Improved physical infrastructure	35
			More efficient operating	20
			Modal shift to lower carbon forms of travel	20
			Low carbon fuels	15
			'Eco-driving' on all modes	10
Total	19.6 mtCO$_2$			

Source: Based on data from Mayor of London, 2007.

Case study: Tokyo

Although Tokyo is among those with the smallest per capita emissions of the developed megacities of the world (among the top five cities in OECD countries; see Table 7.1), its emissions continue to rise. In order to tackle this problem, the Tokyo Metropolitan Government (TMG) aims to create 'Carbon Minus Tokyo', which will cover the following:

- A new look at how energy should be used in cities results in a shift toward a low CO$_2$ society – a low energy society – that allows people to lead an affluent, comfortable urban life while spending the minimum required amount of energy. Low CO$_2$ social systems and technologies that make this society possible should become widespread throughout Tokyo's urban society, thus minimising the greenhouse gas emissions from the metropolis.
- While the optimum use of energy in a manner befitting the characteristics of demand progresses, renewable energies such as solar energy and unutilised energy from urban waste heat should increasingly be put to effective use, thereby enhancing Tokyoites' independence in terms of energy.
- Progress should be made in the passive use of energy that uses natural light, wind and heat as they are, particularly in homes, and the city architecture that considers not only the performance of a building but also the relationship between buildings, structures and greenery around them, and local microclimate.
- The development and subsequent spread of low CO$_2$ social systems and technologies should create a new urban-style business. These social systems, technologies and lifestyles that minimise the environmental burden should enhance the charm of Tokyo as a city, which will be a trailblazing city model that continues to be chosen by people and business enterprises in competition among cities across the world.

In order to achieve its carbon emission reduction objectives, Tokyo proposes to carry out five broad initiatives and several strategies under each of these initiatives.

Initiative I: promote private enterprises' efforts to achieve CO$_2$ reductions

CO$_2$ emissions resulting from corporate activities in the business and industrial sectors in Tokyo account for more than 40 per cent of the city's total emissions, and thus stepping

Table 7.8 Tokyo's emission profile, 1990–2005

		Emissions (mt CO$_2$e)			Growth from base year		Growth from previous year	
		Base year	FY 2004	FY 2005	Growth rate (%)	Growth amount (mt CO$_2$)	Growth rate (%)	Growth amount (mt CO$_2$)
CO$_2$	Industrial	9.9	5.4	5.6	−43.4	−4.3	3.2	0.2
	Business	15.8	20.2	21.0	33.0	5.2	3.9	0.8
	Residential	13.0	14.2	15.0	15.3	2.0	6.2	0.9
	Transport	17.9	20.1	19.3	7.7	1.4	−4.0	−0.8
	Other	1.0	1.0	1.0	−0.9	0.0	1.3	0.0
	Total	57.6	60.8	61.8	7.4	4.3	1.7	1.0
Total for GHGs other than CO$_2$		3.4	2.3	2.2	−36.4	−1.3	−5.6	−0.1
All GHGs		61.0	63.1	64.0	5.0	3.0	1.5	0.9

Source: Based on data from TMG, 2007.

up measures in these sectors is crucial in achieving reductions in the total emissions of greenhouse gases in the metropolis.

- Introduce a cap and trade system targeting large CO$_2$ emitting business establishments.
- Promote smaller businesses' energy conservation measures through the introduction of the Environmental Collateralized Bond Obligation (CBO) Programme.
- Call upon financial institutions to expand environmental investment and loan options and disclose information about investments.
- Achieve the widespread use of renewable energies by promoting the Green Power Purchasing Programme.
- Collaboration in conjunction with smoke, soot and air pollution control measures.

Initiative II: achieve CO$_2$ reductions in households in earnest – cut down on light and fuel expenses by low CO$_2$ lifestyles

Newly built houses in the Tokyo metropolitan area that meet the next generation of energy conservation standards account for no more than 14 per cent of the total, and this level represents less than half the national average. For the widespread introduction of low energy houses, an effort will be promoted in cooperation with housing manufacturers and facility manufacturers in order to raise this ratio of achievement to 65 per cent or so by 2015. Furthermore appliances account for nearly 50 per cent of household energy use. In addition to the on-going energy efficiency labelling system for home appliances, this initiative will:

- wage a campaign for the elimination of incandescent lamps from households;
- build comfortable houses using natural light, heat and wind, and regenerate the solar thermal market;
- improve the energy saving performance of houses (from the current penetration level of 14 per cent of the stock to 65 per cent by 2015);
- facilitate the spread of renewable energies and energy saving equipment such as photovoltaic power generation systems and high efficiency water heaters in houses.

Initiative III: lay down rules for CO_2 reductions in the urban development

- Formulate the world's highest level energy conservation specifications for buildings and apply them to facilities of the TMG:

  - Apply the Tokyo Energy Conservation Design Specifications 2007 to TMG facilities.
  - Formulate guidelines for energy conservation and the introduction of renewable energies to TMG facilities.

- Require large new buildings to have energy conservation performance.
- Introduce an energy conservation performance certificate programme for large new buildings.
- Promote the effective utilisation of energy and the use of renewable energies in local areas.

Initiative IV: accelerate the effort to reduce CO_2 from vehicle traffic

- Formulate rules for the use of fuel efficient vehicles to facilitate the widespread diffusion of hybrid cars.
- Implement a project to encourage the introduction of green vehicle fuel conducive to CO_2 reductions.
- Create a mechanism of support for voluntary activities such as an eco-drive campaign.
- Carry out traffic volume measures by taking advantage of the world's most refined public transportation facilities.

Initiative V: create TMG's own mechanism to support activities in the respective sectors

- Introduce the CO_2 Emission Trading System.
- Create a programme to encourage and support smaller businesses' and households' energy saving efforts.
- Commence a study in terms of tax reduction and taxation to introduce TMG's own energy conservation tax incentive, with a study to be conducted by the Tokyo Metropolitan Tax Research Council.

Case study: New York

With a projected New York municipal population (as opposed to the New York–New Jersey urban agglomeration) of more than 9 million by 2030, New York City (NYC) recognises that its infrastructure is at its 'limits of inheritance'. With ridership at its highest levels in half a century, subways are increasingly jammed; bridges, some over 100 years old, are in need of repair, or even replacement; and the water system, continuously operating since it was first turned on, is leaking and in need of maintenance. The city's energy grid, built with the technology and demand assumptions of an earlier era, strains to meet modern need. Additionally, climate change poses acute risks: by 2030, average temperatures could rise by as much as 2°C in NYC. Hotter temperatures will increase public health risks, particularly for vulnerable populations such as the elderly, and place further strains on infrastructure. The urban heat island could superimpose up to 4°C warming over that caused by global warming. As a city with 830 kilometres of coastline, New York is also at risk of increased flooding as sea levels rise and storms become more intense. The sea levels have already risen 0.3 metres in the last 100 years and are projected to rise by up to 0.25 metres more in the next two decades.

Table 7.9 New York City CO_2 emissions in 2009

Sector	Sub-sector	Current emission (mtCO₂e)	Contribution to the city's total (%)
Buildings		**38.1**	**75**
	Residential	17.27	34
	Commercial	13.21	26
	Industrial	3.56	7
	Institutional	4.06	8
Transportation		**10.1**	**20**
	On-road	8.63	17
	Transit	1.52	3
Solid waste, waste		**2.5**	**5**
	Solid waste, waste water and fugitive	2.50	5
Street lights and traffic signals		**0.1**	**0.2**
	Street lights and traffic signals	0.1	0.2
	Total	**50.8**	

Source: Based on data from PlaNYC, 2011.

NYC sees the challenge of climate change as twofold: the need to reduce its contribution to global warming and preparation for its inevitable effects. Although NYC has among the lowest per capita GHG emissions by US standards (one-third the US average, owing to high density and reliance on mass transit), it recognises the need to reduce its emission further. As a result NYC formulated PlaNYC: A Greener, Greater New York in 2007, with a goal to reduce GHG emissions by more than 30 per cent by 2030 compared to 2005 levels.

Energy use in the built environment of New York is significant. Buildings are responsible for 75 per cent of NYC's carbon emissions, 94 per cent of the electricity use and 85 per cent of potable water consumption.

The plan focuses on the following thematic areas: efficient buildings; clean energy supply; sustainable transportation; and waste management. These themes are further detailed into nine sub-areas:

- *Housing and neighbourhoods:* Create homes for almost a million more city dwellers, while making housing and neighbourhoods more affordable and sustainable.
- *Parks and public space:* Ensure all residents live within a ten-minute walk of a park.
- *Brownfields:* Clean up all contaminated land in NYC.
- *Waterways:* Improve the quality of waterways to increase opportunities for recreation and restore coastal ecosystems.
- *Water supply:* Ensure the high quality and reliability of the water supply system.
- *Transportation:* Expand sustainable transportation choices and ensure the reliability and high quality of the transportation network.
- *Energy:* Reduce energy consumption and make energy systems cleaner and more reliable.
- *Air quality:* Achieve the cleanest air quality of any big US city.
- *Solid waste:* Divert 75 per cent of solid waste from landfills.

Housing and neighbourhoods

- *Create capacity for new housing:*

  o Continue transit oriented re-zonings.
  o Explore additional areas for new development.
  o Enable new and expanded housing models to serve evolving population needs.

- *Finance and facilitate new housing:*

  o Develop new neighbourhoods on underutilised sites.
  o Create new units in existing neighbourhoods.
  o Develop new housing units on existing city properties.

- *Encourage sustainable neighbourhoods:*

  o Foster the creation of greener, greater communities.
  o Increase the sustainability of city financed and public housing.
  o Promote walkable destinations for retail and other services.
  o Preserve and upgrade existing affordable housing.
  o Proactively protect the quality of neighbourhoods and housing.

Parks and public space

- *Target high impact projects in neighbourhoods underserved by parks:*

  o Create tools to identify parks and public space priority areas.
  o Open underutilised spaces as playgrounds or part-time public spaces.
  o Facilitate urban agriculture and community gardening.
  o Continue to expand usable hours at existing sites.

- *Create destination level spaces for all types of recreation:*

  o Create and upgrade flagship parks.
  o Convert former landfills into public space and parkland.
  o Increase opportunities for water based recreation.

- *Re-imagine the public realm:*

  o Activate the streetscape.
  o Improve collaboration between city, state and federal partners.
  o Create a network of green corridors.

- *Promote and protect nature:*

  o Plant 1 million trees.
  o Conserve natural areas.
  o Support ecological connectivity.

- *Ensure the long term health of parks and public space:*

  o Support and encourage stewardship.
  o Incorporate sustainability through the design and maintenance of all public space.

Brownfields

- *Develop programmes to accelerate brownfield clean-up and redevelopment:*

  o Increase participation in the NYC Brownfield Clean-Up Program by partnering with lenders and insurers.
  o Increase the capacity of small businesses and small and mid-size developers to conduct brownfield clean-up and redevelopment.
  o Enable the identification, clean-up and redevelopment of brownfields.
  o Build upon existing state and federal collaborations to improve the city's brownfield programmes.

- *Strengthen incentives for brownfield clean-up and redevelopment:*

  o Study the economic value of brownfield redevelopment in NYC.
  o Use the NYC Brownfield Clean-Up Program to establish funding and other incentives for clean-up and redevelopment.

- *Deepen the commitment to communities for community brownfield planning, education and service:*

  o Support community led planning efforts.
  o Support local and area-wide community brownfield planning efforts.
  o Increase the transparency and accessibility of brownfield clean-up plans.

- *Expand the use of green remediation:*

  o Promote green remediation in the NYC Brownfield Clean-Up Program.
  o Promote green space on remediated brownfield properties.

Waterways

- *Continue implementing grey infrastructure upgrades:*

  ○ Upgrade wastewater treatment plants to achieve secondary treatment standards.
  ○ Upgrade treatment plants to reduce nitrogen discharges.
  ○ Complete cost-effective grey infrastructure projects to reduce combined sewer overflows (CSOs) and improve water quality.
  ○ Expand the sewer network.
  ○ Optimise the existing sewer system.

- *Use green infrastructure to manage storm-water:*

  ○ Expand the Blue-Belt Program.
  ○ Build public green infrastructure projects.
  ○ Engage and enlist communities in sustainable storm-water management.
  ○ Modify codes to increase the capture of storm-water.
  ○ Provide incentives for green infrastructure.

- *Remove industrial pollution from waterways:*

  ○ Actively participate in waterway clean-up efforts.

- *Protect and restore wetlands, aquatic systems and ecological habitat:*

  ○ Enhance wetlands protection.
  ○ Restore and create wetlands.
  ○ Improve wetlands mitigation.
  ○ Improve habitat for aquatic species.

Water supply

- *Ensure the quality of the drinking water:*

  ○ Continue the Watershed Protection Program.
  ○ Protect the water supply from hydro-fracking for natural gas.
  ○ Complete the Catskill/Delaware Ultraviolet (UV) Disinfection Facility.
  ○ Complete the Croton Water Filtration Plant.

- *Maintain and enhance the infrastructure that delivers water to NYC:*

  ○ Repair the Delaware Aqueduct.
  ○ Connect the Delaware and Catskill Aqueducts.
  ○ Pressurise the Catskill Aqueduct.
  ○ Maintain and upgrade dams.

- *Modernise in-city distribution:*

  ○ Complete City Water Tunnel no. 3.
  ○ Build a back-up tunnel to Staten Island.
  ○ Upgrade water main infrastructure.

- *Improve the efficiency of the water supply system:*

  o   Increase operational efficiency with new technology.
  o   Increase water conservation.

Transportation

- *Improve and expand the sustainable transportation infrastructure and options:*

  o   Improve and expand the bus service throughout the city.
  o   Improve and expand the subway and commuter rail.
  o   Expand for-hire vehicle service throughout neighbourhoods.
  o   Promote car sharing.
  o   Expand and improve the ferry service.
  o   Make bicycling safer and more convenient.
  o   Enhance pedestrian access and safety.

- *Reduce congestion on roads and bridges and at airports:*

  o   Pilot technology and pricing based mechanisms to reduce traffic congestion.
  o   Modify parking regulations to balance the needs of neighbourhoods.
  o   Reduce truck congestion on city streets.
  o   Improve freight movement.
  o   Improve gateways to the nation and the world.

- *Maintain and improve the physical condition of roads and the transit system:*

  o   Seek funding to maintain and improve the mass transit network.
  o   Maintain and improve roads and bridges.

Energy

- *Improve energy planning:*

  o   Increase planning and coordination to promote clean, reliable and affordable energy.
  o   Increase energy efficiency.
  o   Implement the Greener, Greater Buildings Plan.
  o   Improve codes and regulations to increase the sustainability of buildings.
  o   Improve compliance with the energy code and track green building improvements city-wide.
  o   Improve energy efficiency in smaller buildings.
  o   Improve energy efficiency in historic buildings.
  o   Provide energy efficiency financing and information.
  o   Create a twenty-first-century energy efficiency workforce.
  o   Make NYC a knowledge centre for energy efficiency and emerging energy strategies.
  o   Provide energy efficiency leadership in city government buildings and operations.
  o   Expand the Mayor's Carbon Challenge to new sectors.

- *Provide cleaner, more reliable and affordable energy:*
  - Support cost-effective repowering or replacement of the most inefficient and costly in-city power plants.
  - Encourage the development of clean distributed generation.
  - Foster the market for renewable energy in NYC.

- *Modernise the transmission and distribution systems:*
  - Increase natural gas transmission and distribution capacity to improve reliability and encourage conversion from highly polluting fuels.
  - Ensure the reliability of NYC power delivery.
  - Develop a smarter and cleaner electric utility grid for NYC.

Air quality

- *Understand the scope of the challenge:*
  - Monitor and model neighbourhood level air quality.

- *Reduce transportation emissions:*
  - Reduce, replace, retrofit and refuel vehicles.
  - Facilitate the adoption of electric vehicles.
  - Reduce emissions from taxis, black cars and for-hire vehicles.
  - Reduce illegal idling.
  - Retrofit ferries and promote the use of cleaner fuels.
  - Work with the Port Authority to implement the Clean Air Strategy for the Port of New York and New Jersey.

- *Reduce emissions from buildings:*
  - Promote the use of cleaner-burning heating fuels.

- *Update codes and standards:*
  - Update codes and regulations to improve indoor air quality.
  - Update the air quality code.

Solid waste

- *Reduce waste by not generating it:*
  - Promote waste prevention opportunities.
  - Increase the reuse of materials.

- *Increase the recovery of resources from the waste stream:*
  - Incentivise recycling.
  - Improve the convenience and ease of recycling.
  - Revise city codes and regulations to reduce construction and demolition waste.
  - Create additional opportunities to recover organic material.
  - Identify additional markets for recycled materials.
  - Pilot conversion technologies.

- *Improve the efficiency of the waste management system:*
    - Reduce the impact of the waste system on communities.
    - Improve commercial solid waste management data.
    - Remove toxic materials from the general waste stream.
- *Reduce the city government's solid waste footprint:*
    - Revise city government procurement practices.
    - Improve the city government's diversion rate.

Table 7.10 Projected reductions of PLANYC by 2030 (in $mtCO_2e$)

Baseline emission (2005)	Business-as-usual change to 2030	Efficient buildings		Clean energy supply		Sustainable transportation		Waste management		Target emission in 2030 (expected)
		Ach.	Proj.	Ach.	Proj.	Ach.	Proj.	Ach.	Proj.	
58.2	+11.7	−1.5	−12.7	−5.9	−4.3	−0.2	−3.1	−1.6	−1.7	40.8 (38.9)

Source: PlaNYC, 2011.

Note: Ach. = reductions achieved to date (2010); Proj. = projected reductions from current and proposed initiatives.

Progress to date

A 30 per cent reduction over the base year (2005) emission of 58.2 million tons of CO_2e in 2030 yields a target of 40.8 million tons of CO_2e. The current and planned reductions are expected to yield an emission of 38.9 million tons of CO_2e in 2030.

In just four years (2006–10), NYC has created or preserved over 64,000 units of housing, and completed over 20 transit oriented re-zonings so that more than 87 per cent of new development is transit accessible. NYC has embarked on a new era of parks construction, bringing over 250,000 more residents within a ten-minute walk of a park. The city's first bus rapid transit system has been launched, and $1.5 billion investment has been committed for green infrastructure to clean the waterways. Additionally nearly half a million trees have been planted, and investment in the drinking water supply network has been made.

Over 30 per cent of the yellow taxi fleet is now 'green', reducing emissions from some of the most heavily used vehicles. The city has enacted regulations to phase out dirty heating fuels, which are responsible for more pollution than all of the cars and trucks on NYC streets. The process to remediate brownfields has been streamlined, reducing the average time it takes to begin a clean-up of the city's most polluted plots. Public plazas for pedestrians include one in Times Square, the 'crossroads of the world', and they are attracting tourists and New Yorkers alike. Pedestrian fatalities are down. NYC has completed over 100 energy efficiency retrofits on city owned buildings as part of the commitment to reduce city government greenhouse gas emissions 30 per cent by 2017. Landmark green building legislation has been enacted.

At the same time, many obstacles to achieving some of the PlaNYC goals remain. Efforts to maintain, improve and expand the transit network have been stymied by the lack of a stable, sufficient and rational funding source. Congestion continues to clog NYC streets; the global recession has forced the city to reduce its capital budget; and some PlaNYC projects have been delayed. Several initiatives have also been slowed by a lack of state or federal permission, action or funding.

7.4 Carbon management in developing cities

Global urbanisation, at its current peak, is largely a phenomenon of the developing world. By 2030, the global urban population will be nearly 70 per cent of the total population, and 80 per cent of urban humanity will live in the developing world (UNPD, 2010). While the causes for carbon emissions in cities are similar in all cities, important differences exist between cities of the developed and the developing world. Many people in the developing world continue to see cities as places for opportunity, convenience, culture and indeed a new lifestyle. The attraction of cities of the developing world to their citizenry continues, despite acute problems of overcrowding, lack of sanitation, air pollution and urban warming.

Lebel *et al.* (2007) suggested that developing cities have four core functions that people find attractive: mobility, shelter, food and lifestyle. These functions are affected by the urban form (as in the case of all cities) as well as the type of city (whether it is a centre of government, industry, service or education), all of which have implications for a city's GHG emissions (Figure 7.2).

7.4.1 Mobility

The trend in developing cities is still mostly away from self-employed, home based enterprises to employment in firms elsewhere. The prospects of telecommuting, often muted for post-industrial societies (e.g. Tayyaran and Khan, 2003), remain therefore remote. Moreover, Lebel *et al.* (2007) suggest scepticism, as people's choices and use of automobiles, for example, may have less to do with 'going to work' than with needs to make business and shopping trips, which can be expected to expand with increasing wealth.

A key goal must therefore be to introduce less carbon intensive mobility systems. Well-designed, multimodal systems organised around public transit could help shape urban form rather than just respond to it. Much more investment needs to be put into public mass transit systems rather than roads for private vehicles. Multilateral financial institutions have a major responsibility to reorient their loan priorities away from conventional road building toward financing mass transit systems. A combined carrot and stick approach to private car ownership and use (taxes)

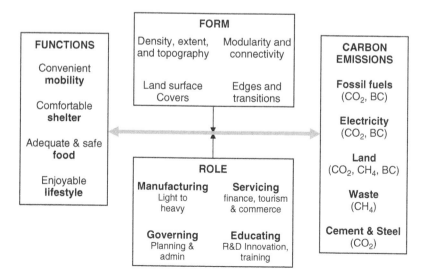

Figure 7.2   Core functions of a developing city and its emission implications

Table 7.11 Challenges and opportunities for decoupling urban growth from carbon emissions in developing cities

Urban service	Critical links to emission	Key carbon challenges	Opportunities to integrate carbon into development strategies
Mobility	Motorised transport – CO_2 and particulate matter (PM10 and PM2.5).	Rise in personal vehicle ownership. Weak urban form. Poor transport service provision.	*De-motorisation:* High density and greater use can make mass transit feasible.
Shelter	Embodied emission in electricity and building materials, especially cement and steel. Operational energy demand to cool interiors.	Greater demand for air conditioning. Energy inefficient modes of construction.	*Climate sensitive design:* Passive design to enhance human comfort and health. Mitigation of UHIs. Appropriate selection of construction materials.
Food	Methane from livestock. Carbon from clearing forest for agriculture.	Increased consumption of high energy food (meat and dairy products).	*Healthy and adequate:* Protein substitute. Efficient production and processing with waste recycling and energy capture.
Lifestyle	Energy consumed and other pollutants emitted in manufacturing. Indirect and deemed emission in service sector work.	Overconsumption and associated waste from consumables. Poor regulations and/or perverse subsidies leading to high pollution intensities.	*Modest footprints:* Low energy and material efficient goods and services. Wise use of information and communication technologies.

Source: Derived from Lebel *et al.*, 2007; Emmanuel, 2005.

combined with good provision of public transport could work well in developing cities. To make these strategies work, Lebel *et al.* (2007) suggested that core retail, education and entertainment areas might be closed permanently to above-ground motorised transport.

Urban land use and transport infrastructure will need to be much better coordinated, whether this is through planning and regulation or more self-organising opportunities to re-zone and reconfigure.

7.4.2 *Shelter*

Shelter in developing cities has to be provided within the constraints of high density, increasing urbanisation, and altered microclimate. In the face of rapid urbanisation and shrinking resources to cater to the urban needs, higher densities of urban growth are required of all developing cities. That they should cool themselves in tropical regions by passive means in an ecologically sensible manner goes without saying. At the same time, the problem of housing density and growing human conglomerations such as Djakarta, Manila, Accra and Cartagena is not so much the lack of resources or space, but one of rational allocation (Emmanuel, 2005). If only the designers could calculate the actual intensity of current development patterns in these cities, it would not be long before they discovered that much denser growth is possible. Two new ways of looking at urban growth must however be accepted before such conclusions can be put to practical use.

The apparently high building density in the tropics is not so much due to high rise buildings, but because the occupancy rate per room is high (Correa, 1989: 42). Every available space is built upon illegally by urban migrants. There is a crying need for open space in tropical cities. The subdividing of city land partitions even the little open space available into private, but thermally and ecologically insignificant, backyards. While open space in a developed city like London is 3 hectares per 1,000 people, Delhi has only 1.5 hectares per 1,000 people. A developing city like Mumbai has only 0.1 hectare. This paltry amount includes even the grass on traffic islands (Correa, 1989: 42).

Tay Kheng Soon (2001) argued that higher densities in the tropics are theoretically possible, provided common amenities are adequately provided and access to common areas is ensured. Herein lies an important conceptual direction for climate-conscious urban design in the tropics: the design of spaces between buildings in such a climatically suitable way as to ensure usability and accessibility.

In a landmark study conducted in temperate cities, March and Martin (1972) found that even the apparently dense North American cities have very low floor area ratios (FAR). They also showed that a city like New York could be developed to its present intensity with much smaller buildings (not more than six to eight floors) if the road networks could be re-organised. Such re-organisation can be the second conceptual direction for high density, low rise, climate-conscious growth in the tropics: a re-thinking of urban transportation.

Many European cities are in the process of re-discovering the possibilities of high density, low energy, more humane urban growth brought about by pedestrianisation. Cities like Amsterdam, Stockholm, Athens (Havlick, 1983) and Copenhagen (Gehl, 1989) have re-organised city street networks so as to facilitate a climatically and aesthetically appropriate growth. Even a motorised city like New York depends for a large part on pedestrianisation. It is estimated that more than 70 per cent of all rush-hour traffic in Manhattan is on foot (Wright, 1992). The tautology of mass transit and density ('Mass transit will not work unless density is high: People will not live densely until a good mass transit system is established') can be solved only by moving in the direction of high density development and pedestrianisation simultaneously.

Once these two pre-conditions (design of accessible public spaces and pedestrianisation) are met, a new approach to climate-conscious design in the urban tropics can be attempted. In other words, the spaces between buildings must be designed as sensitive to climate, and the use of such spaces must be made part and parcel of daily urban living, by ensuring accessibility. Emmanuel (2005) termed this the design of the 'commons'.

7.4.3 Design of the 'commons'

The idea of the 'commons' was first developed as a way of understanding human population dynamics by William Forster Lloyd in 1833 (Hardin and Badden, 1977). It was later developed as a notion of that which is opposed to the 'private'. In our age, 'commons' refers to 'that part of the environment that lay beyond a person's own threshold and outside his own possession, but to which, however, that person had a recognised claim of usage – not to produce commodities but to provide for the subsistence of kin' (Illich, 1982: 16). Illich's definition includes the pavements and semi-public areas of street where one can relate to other human beings. 'Commons' is taken to mean all realities that are naturally endowed for the common good of all. In this sense, it would include physical attributes like meadows, scenic views, beaches and so on, and psycho-social attributes like common language, culture, caring for the elderly and sick, and so on. In this chapter we will limit the notion of the commons to its spatial attributes only. Illich clarifies this understanding better:

> Just as the home reflects in its shape the rhythm and extent of family life, so the commons are the traces of the commonality. There can be no dwelling without its commons. . . . Space fit to bear the marks of life is as basic for survival as clean water and fresh air. Human beings simply do not fit into garages, no matter how splendidly furnished with . . . . energy saving devices.
>
> (Illich, 1982: 60)

One of the problems with past attempts at climate-conscious design was the insensitivity to the qualitative attributes of design, especially at the urban scale. Energy saving techniques are expected to be widely used simply because of their economic potential. Experience has shown that such assumptions do not always hold. The treatment of climate sensitive design as an issue concerned with the commons will help humanise them. Furthermore, it is only at such scales (between buildings and neighbourhoods) that climate-conscious design can effectively tackle the problems caused by urban heat islands while ensuring high density living.

However, there are certain urban activities that best suit the commons and others that work only in the private. Correa (1989) identifies a hierarchy of four urban space usage patterns that are typical of the developing world: space needed by the family for private uses (cooking, sleeping, storage and so on), intimate contact space with the outside world (the front door step), neighbourhood meeting places (wells, city water taps) and urban gathering places (the maidan). The last three fall within the domain of the commons to varying degrees. Climate-conscious urban design initiatives in the urban tropics must concern themselves with these three kinds of urban space needs.

Tropical living is a part indoors and part outdoors activity, though one suspects mostly the latter. Living in the tropical outdoors is relatively pleasant for most of the year. Correa quantifies this aspect of flexibility of outdoor use by a factor called the 'usability co-efficient' (Correa, 1989). While people spend a longer time at any given time in the indoors, outdoor occupants are typically there for shorter periods at a given time. Thus shading to reduce the thermal stress and perceivable air movement to enhance the cooling effect are necessary in the outdoors.

A sole reliance on wind movement alone is not tenable in the urban tropics. Although such a strategy is routinely touted as a panacea for all climatic ills in the tropics, we must remember two facts about tropical wind patterns. On the one hand, the night-time rate of air movement is low or nil in the tropics. Furthermore, being in the 'doldrums' twice a year for extensive periods, the equatorial tropics do not have high wind velocities at the macro-level. Most daily wind flows are local in nature (like the sea–land breeze). A nocturnal urban heat island tends to keep the air temperature over land almost as high as that over sea, thus weakening any night-time land breeze.

The 'problem' of climate-conscious tropical urban design is therefore twofold: prevention of heat build-up as the day unfolds (so as to reduce night-time cooling loads); and encouraging convective cooling at night.

Viewed in this manner, the urban design goals would be a) radiation reduction during the day and b) ventilative cooling at night. The nature of the climatic problem is such that neither of these goals can be considered in isolation.

7.4.4 Food

Cities and urbanising regions are points of convergence of many production–consumption systems. They provide opportunities to influence both deemed and direct emissions, for example through regulations, policy integration and technological upgrading and by shaping norms and consumer cultures. This is clearly the case in food. Diets change with urbanisation and contribute to changes in land use and in supporting landscapes. The implications for carbon sequestration and emissions of intensified production practices, new consumption patterns, redistribution of the rural workforce, and competition between urban and rural land and water use are complex (Lebel *et al.*, 2007).

7.4.5 Lifestyle

Spatial planning is critical and can build on the legacies of market and other social spaces or innovations in decentralised local government. Fun and meaningful cultural activities need not be carbon intensive. A good example is street closures to create new inner urban social spaces for eating and meeting. Local performances of music and plays in the tropical climate can take place out of doors in the evenings with modest energy requirements. Higher densities mean that better use of public and private space may be warranted, for example through sharing and multiple uses of venues on weekdays and at weekends.

Case study: Bangkok

The Bangkok Metropolitan Area (BMA) is a primate city region in Thailand, consuming approximately 29,200 GWh of electricity annually, which is equivalent to 14.86 million tons of CO_2 emissions (34 per cent of total emission in BMA). Transportation consumes approximately 28 million litres of gasoline per day (approximately 21.18 million tons of CO_2 or 50 per cent of the total emissions). Methane from solid waste landfill and wastewater is estimated at 1.13 million tons of CO_2 equivalent (or 3 per cent of total emissions). Other activities such as rice fields and canals produce a further 5.58 million tons of CO_2 equivalent annually (or 13 per cent of the total).

BMA's Action Plan on Global Warming Mitigation (BMA, 2007) aims to reduce the total emission by 15 per cent over five years, using the following strategies:

1   Reduce energy consumption and maximise efficiencies in resource utilisation in all activities to minimise global impacts.
2   Promote and support all sectors and stakeholders to reduce GHG emissions jointly.
3   Promote the sufficiency economy lifestyle to prepare for, and adapt to, global warming.
4   Promote and support activities that lead to GHG absorption.
5   Promote and support activities that continuously work to mitigate global warming by building public awareness and knowledge (BMA, 2007).

These strategies are translated into five initiatives in the action plan and were adopted in May 2007:

• *Initiative 1:* Expand mass transit and improve traffic systems.
• *Initiative 2:* Promote the use of renewable energy.
• *Initiative 3:* Improve electricity consumption efficiency.
• *Initiative 4:* Improve solid waste management and wastewater treatment efficiency.
• *Initiative 5:* Expand park areas.

Initiative 1: Expand mass transit and improve traffic systems

• *Action plan 1:* Expand the mass transit rail system within the BMA.
• *Action plan 2:* Improve the public bus system.
• *Action plan 3:* Improve the traffic system.

Other supporting activities include:

- Develop more park and ride facilities to support passenger car drivers' use of mass transit when travelling to the inner city area.
- Build more bike lanes to encourage greater use of bicycles.
- Implement and promote a common ticket system for public transit users in Bangkok.

Initiative 2: Promote the use of renewable energy

- *Action plan 1:* Promote the use of biofuels.

Other supporting activities include:

- Campaign for the use of low carbon emission petroleum fuel or use of liquefied gas, e.g. CNG.
- The Bangkok Metropolitan Administration to facilitate the buying and collecting of used cooking oil for refining bio-diesel.

Initiative 3: Improve building electricity consumption efficiency

- *Action plan 1:* Improve building energy consumption efficiency (improve the energy efficiency of Bangkok Metropolitan Administration's buildings, and promote and support the implementation of energy conservation schemes in privately owned buildings).
- *Action plan 2:* Promote an electricity conservation campaign for Bangkokians (campaign for efficient use of electrical appliances, campaign for reduced use of air conditioning, support energy efficiency labelling of, and proper maintenance schemes for, electrical appliances, promote the use of energy saving appliances, and promote the use of energy saving light bulbs).

Initiative 4: Improve solid waste management and wastewater treatment efficiency

- *Action plan 1:* Increase efficiency in solid waste management (improve efficiency in organic waste management, and support solid waste reuse and recycling).
- *Action plan 2:* Increase efficiency in wastewater treatment (increase wastewater treatment capacity, and reduce household wastewater).

Initiative 5: Expand park areas

- *Action plan 1:* Plant trees in the BMA.
- *Action plan 2:* Plant trees in the neighbouring province areas.

These initiatives and actions are expected to reduce GHG emission by 9.75 million tons of CO_2 equivalent (see Table 7.12 for details).

Table 7.12 Bangkok GHG emission reduction plan, 2007–12

Sector	Baseline GHG emission (mt CO$_2$e)	Business-as-usual GHG emission (2012) (mt CO$_2$e)	Actions in 2007–12 to reduce emissions	Expected reduction (mtCO$_2$e)	Emission in 2012 under BMA Action Plan (mtCO$_2$e)
Transportation	21.18	25.30	Mass transit rail	−2.40	
			Bus rapid transit (BRT) system	−0.19	
			Improve existing bus network	−1.24	
			Improve road network	−1.70	
			Total reductions	*−5.53*	19.77
Energy production			Promote use of gasohol	−0.27	
			Promote use of bio-diesel	−0.34	
			Total reductions	*−0.61*	−0.61
Buildings	14.86	16.00	Improve BMA buildings	−0.01	
			Improve private buildings	−0.42	
			Efficient use of electrical appliances	−0.70	
			Reduce A/C use	−0.41	
			Appliance labelling system	−0.44	
			Energy saving appliances	−0.14	
			Energy saving light bulbs	−0.13	
			Total reductions	*−2.25*	13.75
Waste	1.13	1.13	Improve efficiency of organic waste management	−0.10	
			Solid waste reuse and recycle	−0.28	
			Increase wastewater treatment capacity	−0.05	
			Reduce household wastewater	−0.03	
			Total reductions	*−0.46*	0.67
Urban green	−0.10	−0.10	Plant trees in BMA land	−0.14	
			Support private tree planting	−0.32	
			Plant trees in neighbouring provinces	−0.54	
			Total reductions	*−1.00*	−1.00
Others	5.58	6.36			6.36
Total	42.65	48.69		−9.85	38.94

Source: Based on data from BMA, 2007.

Case study: Singapore

Given the specific circumstances of Singapore (a small island city-state, with limited forest cover) it is not possible to implement standard carbon mitigation practices. Coupled with the already well-developed transportation policies to tackle both carbon emission and congestion, Singapore's options to mitigate GHG emissions are largely confined to increasing its energy efficiency and using less carbon intensive fuels. The use of large scale renewables such as biomass is impractical in this small island state, while the use of others such as wind and geothermal energy is technically not feasible. The large scale use of other renewable technologies such as photovoltaics remains largely unproven in tropical climates.

Given these realities, Singapore's climate change action plan aims to:

1   promote the adoption of energy efficient technology and measures by addressing the market barriers to energy efficiency;
2   raise awareness to reach out to the public and businesses so as to stimulate energy efficient behaviour and practices;
3   build capability to drive and sustain energy efficiency efforts and to develop the local knowledge base and expertise in energy management;
4   promote research and development to enhance Singapore's capability in energy efficient technologies.

As a relatively low-lying, densely populated island in the tropics, Singapore is affected by climate change. Much of the island is less than 15 metres above sea level, with a generally flat coast. With a population of about 4.7 million within its 193 kilometre coastline, Singapore is one of the most densely populated countries in the world. In addition, Singapore has a relatively high uniform temperature and abundant rainfall, and is also situated in a region in which communicable diseases such as dengue are endemic (MOEWR, 2007).

Given the paucity of mitigation actions Singapore can take and the vulnerability to climate change Singapore faces as outlined above, Singapore has chosen to focus on adaptation measures as part of its carbon and climate change action. The key areas of action include the prevention of the following:

1   increased flooding;
2   coastal land loss;
3   water resource scarcity;
4   public health impact from resurgence of diseases;
5   heat stress;
6   increased energy demand;
7   impacts on biodiversity.

Table 7.13 lists the key adaptation measures currently undertaken by Singapore. Table 7.14 shows the key GHG mitigation measures proposed by MOEWR (2007).

Table 7.13 Key adaptation measures proposed by Singapore

Climate change threat	Key adaptation measures
Flooding	Reduce flood prone areas from 124 hectares in 2007 to less than 66 hectares by 2011.
	Complete the Marina Barge Project to alleviate flooding, increase water supply and enhance local quality of life.
	All reclamation projects to be 125 centimetres above the highest recorded tide level.
Coastal erosion	Increase the resilience of the already heavily protected coastline (70–80% of the Singapore coastline is hardlined).
Water scarcity	Diversify water supply by NEWater and desalination projects.
	Marina Barge Project (see above, under 'Flooding').

Heat stress	Promote heat island mitigation efforts:
	Increase greenery in the city (e.g. city parks, rooftop gardens, vertical greening in buildings).
	Modify building layouts and designs (e.g. using building materials with better thermal properties, lighter coloured building surfaces).
	Design building interiors and exterior building layouts for better ventilation. Maximise the wind tunnel effect.
Higher energy demand due to air conditioning	Improve building energy efficiency by:
	A high performance building envelope that meets the prescribed envelope thermal transfer value (ETTV), currently set at 50W/m².
	Explore the possibility of extending the ETTV regulations to residential buildings.
	Stipulate minimum performance standards ('Green Mark') for new buildings.
Resurgence of diseases	Comprehensive mosquito surveillance, control and enforcement system to minimise vector borne diseases:
	Pre-emptive action to suppress the mosquito vector population.
	Dengue-related research.
	Review of building design to reduce potential breeding habitats (e.g. roof gutters in new buildings have been prohibited except in special circumstances).
Loss of island and marine biodiversity	Monitoring long term tree diversity, tree growth and survival in marked study plots. Coral nursery off Palau Semakau.
	Pre-emptive management strategies to counter mangrove erosion at some coastal areas.

Source: Based on MOEWR, 2007.

Table 7.14 Key policy measures to mitigate GHG emissions in Singapore

	Power generation	Industry	Buildings	Transport	Households
Promote adoption of energy efficient technology and measures	Clean development mechanism				
	$10 million EASe Scheme				
	Accelerated depreciation allowance				
	Investment allowance				
	Promote cogeneration and trigene-ration via industrial land planning and facility siting	Design for Efficiency Scheme Grant for energy efficient technologies	Building regulations Government takes the lead Energy Smart Mandating Green Mark certified $20 million Green Mark Incentive Scheme	Manage vehicle usage and traffic congestion Improving and promoting the use of public transport Fuel economy labelling Green vehicle rebate	Mandatory labelling Minimum energy performance standards Electricity Vending System Electricity consumption tracking device

Table 7.14 Continued

	Power generation	Industry	Buildings	Transport	Households
			Grant to upgrade building envelopes	Promoting fuel efficient driving habits	
			Residential building standards		
Research and development and capacity building	Innovation for Environmental Sustainability Fund				
			Green buildings R&D fund		
	Energy service company accreditation scheme Singapore certified energy manager Programme and training grant				
Raise awareness	Energy efficiency seminars and workshops Energy efficiency website Public awareness programme				

Source: MOEWR, 2007.

7.5 Urban adaptation strategies to manage climate change

Given the current levels of urbanisation and the growth trajectories, much of the consequences of excessive anthropogenic GHG emissions will be faced by urban dwellers. While the preceding discussions highlighted actions cities could take to mitigate their emissions, it is important to remember that the in-built lag effect of climate change is such that the more urgent need is to adapt our lives to climate change. Cities are at the forefront of adapting human lives to climate change, even as they both drive the emissions and develop knowledge, tools and processes to manage the change.

Three aspects of climate change can have the most serious consequences to cities: warming (and associated heatwaves and health hazards); flooding (irregular precipitation combined with infrastructural inadequacies); and sea level rise (and associated coastal erosion) (see Table 7.15). Urban adaptation to climate change in a carbon effective manner will be key not only to manage the change but also to mitigate the emissions in the long run so as to reduce the need for adaptation.

7.5.1 Adapting to urban warming – planning approaches to tackle the urban heat island

Given the relatively small fraction of global land cover under cities, and the magnitude of regionally averaged urban anthropogenic heat generation compared to solar loading, it is highly unlikely that cities directly influence global warming. However, the warm and polluted air plume generated by the UHI effect can alter atmospheric chemistry to such an extent that large scale atmospheric processes are altered (Crutzen, 2004). The impact of urban land cover becomes relatively

Table 7.15 Key climate change adaptation needs in developing cities

Change in climate	Possible impact in developing cities
Changes in means:	
Temperature	Increased energy demands for heating and cooling.
	Worsening of air quality.
	High temperature impacts exaggerated by urban heat islands in cities.
Precipitation	Increased risk of flooding.
	Increased risk of landslides.
	Distress migration from rural areas.
	Interruption of food supply networks.
Sea level rise	Coastal flooding.
	Reduced income from agriculture and tourism.
	Salinisation of water sources.
Changes in extremes:	
Extreme rainfall and	More intense flooding.
tropical cyclones	Higher risk of landslides.
	Disruption to livelihoods and city economies.
	Damage to homes, infrastructure and businesses.
Drought	Water shortages.
	Higher food prices.
	Disruption of hydro-electricity.
	Distress migration from rural areas.
Heat- or cold waves	Short-term increase in energy demands for heating or cooling.
	Health impacts for vulnerable populations.
Abrupt climate change	Possible significant impacts from rapid and extreme sea level rise.
	Possible significant impacts from rapid and extreme temperature change.

Source: Based on Bartlett *et al.*, 2011.

more important as the fraction of urban land cover increases (Lamptey, 2010). Furthermore, different methods of disentangling the urban influence from observed climate changes produce different outcomes. For example, Parker (2010) estimated the likely error in global land surface air temperature due to urban warming to be 0.006°C per decade (and 0.012°C per decade at night). Kalnay and Cai (2003) on the other hand estimated a land cover change effect due to urbanisation and agriculture to be 0.35°C per century.

Analysing the urban influence on Taiwan's climate in the last 50 years, Lai and Cheng (2010) found that rapid expansion in population and economic activities influenced air temperature changes directly (i.e. more energy use leading to higher greenhouse gas emissions) and indirectly (i.e. changing the urban landscape such as the depth of street canyons, vegetation space, amount of sky view, and heat capacity).

For tropical regions, the warm and polluted plume of air rising from cities, together with high Bowen ratios (i.e. sensible relative to latent heating) in the boundary layer, may stimulate deep convection and lightning, particularly during the dry season (Crutzen, 2004). However, Lal and Pawar (2011) found that in India the effect of UHI on deep convection and lightning is more pronounced in cities with less appreciable aerosol pollution (whereas pollution was a greater contributor to these effects in highly polluted cities). The UHI effect may also alter regional circulations such as the sea–land breeze (Lu *et al.*, 2010; Nieuwolt, 1966), which in turn may affect regional pollution dispersion.

Urban design and planning strategies targeting the amelioration of urban warming are rare. It is even rarer to see an explicit link being made with urban warming strategies and the adaptation to global or regional warming as well as the mitigation of carbon emission. Among cities with warm climates, Japanese cities lead the way in using UHI mitigation expressly as a global warming adaptation approach (see for example JaGBC, 2006). Japan's Outline Policy Framework to Reduce UHI Effects adopted in March 2004 (cited by Yamamoto, 2006) aims to:

- reduce anthropogenic heat release (building and plant energy efficiency, building insulation and shade, the greening of buildings, reflective surfaces, district cooling systems, and reuse of waste heat);
- improve artificial surface covers (reflective pavements and water retention, green cover, and open spaces);
- improve urban structures (building morphology and land use manipulation, and greenery) (Yamamoto, 2006).

An empirical evaluation of the effect of urban morphology on local climate in Beijing in summer (baseline daily mean = 24.9°C; daily maximum air temperature = 30.2°C) by Zhao *et al.* (2011) found that three planning indicators (building density as given by floor area ratio, building height, and green cover) can explain nearly 99 per cent of the local microclimate differences (surface temperature, peak temperature, and time of day of occurrence of peak temperature). Given the climatic similarities between the summer conditions in Beijing and central Japan cities and warm, humid cities, it is likely that the manipulation of the following variables could lead to greater reduction in local warming in warm, humid cities.

Shade

Evidence from the warm, humid tropics as well as from cities with warm, humid summer conditions indicates that shade (caused by either buildings or trees) is the single most important design parameter in determining local warming or cooling, as the radiative flux from direct sunlight has a strong influence on the heat balance of the body (Taylor and Guthrie, 2008). Emmanuel and Johansson (2006) showed that shading can be the main strategy for lowering air and radiant temperatures in the warm, humid city of Colombo, Sri Lanka. This can be achieved by more compact urban form with deeper street canyons, covered walkways and shade trees (Johansson and Emmanuel, 2006). In the high density settings of Hong Kong, Yang, Lau and Qian (2010) found that, on a diurnal basis, 'the semi-enclosed plot layout with high density and tree cover has the best outdoor thermal condition'. Similarly, an annual outdoor thermal comfort study in Taiwan (Hwang, Lin and Matzarakis, 2011) found that outdoor thermal comfort is best when a location is shaded during spring, summer and autumn. Whether the shade come from trees or shading devices makes little difference.

Noting the benign neglect of street level shading in contemporary urban planning and design in hot climates, Erell (2008) states that, in hot climates with high radiant loads, net radiant balance may be more important than convective exchange. This has a greater effect on pedestrian comfort than the minor modifications to air temperature usually reported in street level measurements.

In summary, the worst street level comfort conditions in warm, humid regions are associated with wide streets lined with low rise buildings and no shade trees. The most comfortable conditions are associated with narrow streets and tall buildings, especially if shade trees are also present.

Ventilation

Ventilation has been a key strategy for thermal comfort and pollution dispersal in hot climates from ancient times. However, the low levels of wind speeds in the tropics due to the twice-a-year passage of the inter-tropical convergence zone make it necessary to map out carefully the ventilation strategy at a city-wide level to induce sufficient air movement, both for pollution dispersion and for thermal comfort. It will also enhance the cooling potential of naturally ventilated buildings (which remains the commonest approach to indoor cooling in the warm, humid tropics). Hong Kong's approach to a city-wide ventilation strategy via the air ventilation assessment (AVA) method (Ng, 2009) best exemplifies such a planning assessment method.

Ng (2009) suggested that ventilation strategies for pollution dispersal and thermal comfort need to consider both city-wide and street level measures such as breezeways or paths (city-wide scale), building plot coverage, building orientation relative to streets, heights of building in relation to one another, and building permeability. Incorporating such an approach, Chapter 11 of the Hong Kong Planning Standards and Guidelines (http://www.pland.gov.hk/pland_en/tech_doc/hkpsg/full/ch11/ch11_text.htm#8.Implementation) states:

> For better urban air ventilation in a dense, hot-humid city, breezeways along major prevailing wind directions and air paths intersecting the breezeways should be provided in order to allow effective air movements into the urban area to remove heat, gases and particulates and to improve the micro-climate of urban environment.

Another strategy to induce street level ventilation is to use the differences in surface temperatures of vertical surfaces of buildings in high density areas. For example, Yang and Li (2009) found that the surface temperatures of walls in high density Hong Kong increase with height during the day and reverse at night, leading to ventilation induced by thermal buoyancy that could be two to four times stronger than mountain slope flows.

Urban greenery

The importance of urban greenery to human comfort at street level is long recognised. However, efforts to use greenery to ameliorate urban warming need to be cognisant of the scale of the effect due to different kinds of greenery, limitations of its use and the unintended consequences that might arise by its haphazard deployment. Furthermore, the impact of specific greening interventions on the wider urban area, and whether the effects are due to greening alone, has yet to be demonstrated (Bowler *et al.*, 2010).

Urban greenery to tackle local warming in warm, humid cities may take the form of green roofs, green walls, surface green (such as lawns) or street trees. Green roofs have attained high prominence in recent years as a UHI mitigation strategy, yet their effect on reducing 'urban' warming (as opposed to a positive contribution to building energy consumption) remains unclear.

Alexandri and Jones (2008) simulated the effect of green roofs and green walls in different climates (including cities with warm, humid summers such as Hong Kong, Brasilia and Mumbai). They found that humid climates can benefit from green surfaces, especially when both walls and roofs are covered with vegetation. Green walls were shown to have a stronger temperature effect inside street canyons than green roofs under all climates. The cooling effect of green surfaces increased in proportion to the amount surfaces exposed to the sun (Alexandri and Jones, 2008). This further emphasises the importance of shading in the first instance, which could enhance the cooling potential of green surfaces in warm, humid areas.

Although the effects of green cover manipulations on street level air temperature remain muted, their effect on thermal comfort is significant. Spangenberg *et al.* (2009) simulated green cover conditions in São Paulo, Brazil (average summer temperature range 22–30°C) and found that incorporating street trees in the urban canyon had a limited cooling effect on the air temperature (up to 1.1°C), but led to a significant cooling of the street surface (up to 12°C) as well as the mean radiant temperature at pedestrian height (up to 24°C). Although the trees lowered the wind speed up to 45 per cent of the maximum values, the thermal comfort was improved considerably, as the physiologically equivalent temperature (PET) was reduced by up to 12°C. This was further confirmed by Hwang *et al.* (2010) in Taichung, central Taiwan (average summer conditions: mean air temperature 28.5°C; relative humidity range 70–80 per cent), where a completely tree covered outdoor space was found to be acceptable to 80 per cent of the users even at an air temperature of 30°C, where the comfortable operative temperature varied between 24.5 and 32.5°C).

A meta-review conducted by Bowler *et al.* (2010) on the purported effect of urban greenery best sums up the findings to date. Noting the nature of observational studies on the urban green effect (small numbers of green sites), Bowler *et al.* (2010) concluded that 'the impact of specific greening interventions on the wider urban area, and whether the effects are due to greening alone, has yet to be demonstrated'. Further empirical research is necessary in order to guide the design and planning of urban green space efficiently, and specifically to investigate the importance of the abundance, distribution and type of greening. It is also necessary to be mindful of the interference urban greenery could cause to street level pollution removal, especially on the leeward side of urban canyons (Gromke *et al.*, 2008; Salim, Cheah and Chan, 2011), as well as the enhanced water use that might be required to maintain the green cover (Gober *et al.*, 2010).

Albedo

In the typically low wind speeds prevalent in tropical cities, the effect of façade materials and their colours assumes greater significance. Priyadarsini, Wong and Cheong (2008) found that low albedo façade materials in Singapore led to a temperature increase of up to 2.5°C at the middle of a narrow canyon. Emmanuel and Fernando (2007) found that high albedo could make sunlit urban street canyons up to 1.2°C cooler in Colombo, Sri Lanka.

However, it is important to keep in mind that albedo enhancement strategies, such as urban greening, are more likely to show improvements in air temperatures than in thermal comfort (Emmanuel, Johansson and Rosenlund, 2007). From an urban design point of view, mitigation options ought to focus on thermal comfort enhancement (including the MRT) rather than merely attempting to control air temperatures (Emmanuel and Fernando, 2007).

A more promising approach to cool the many dark surfaces in cities that cannot be effectively shaded is the use of the so-called cool materials (either low albedo or phase change materials – PCMs). Akbari and Levinson (2008) presented several approaches to low albedo roofs and related standards in different climatic regions in the USA. Synnefa *et al.* (2011) showed that PCMs can effectively reduce surface temperatures of dark asphalt surfaces in cities (typically these are 'hard to treat'). The added advantage of PCMs is that they are available in many colours, thus eliminating the need for white surfaces (with their attendant maintenance problems in humid environments). These strategies could not only reduce building energy consumption but also lead to city-wide lowering of ambient air temperatures, slowing ozone formation and increasing human comfort.

7.5.2 Adapting to flooding, sea level rise and coastal erosion

Many coastal populations are at risk from flooding, particularly when high tides combine with storm surges and/or high river flows. Between 1994 and 2004, about one-third of the 1,562 flood disasters, half of the 120,000 people killed, and 98 per cent of the 2 million people affected by flood disasters were in Asia, where there are large population agglomerations in the flood plains of major rivers (e.g. the Ganges–Brahmaputra, the Mekong and the Yangtze) and in cyclone prone coastal regions (e.g. the Bay of Bengal, the South China Sea and the Philippines). Eight of the top ten countries with the largest number of people living in low elevation coastal zones are in the developing world. Nine of the ten countries with the largest fraction of their people living in such areas are also in the developing world (McGranahan, Balk and Anderson, 2007). There is a statistically significant difference between the urban population in the low elevation coastal zone (LECZ) for low income versus high income countries. Politically notable is the appreciable difference found between the populations of the least developed countries (LDCs) and OECD countries living in the zone: 14 per cent and 10 per cent respectively. This divide is even wider when looking at the urban population percentages in the LECZ: 21 per cent for the LDCs and 11 per cent for OECD countries.

Urban areas are more prone to flooding on account of increased impervious cover and filling of wetlands and other natural flood retention systems to make way for 'development'. Land compaction and subsidence due to excessive withdrawal of ground water for human consumption not only can lead to flooding but also can increase sea water intrusion in coastal urban areas (see McGranahan *et al.*, 2007).

Effective adaptation will require a combination of effective and enforceable regulations and economic incentives to redirect new settlement to better protected locations and to promote investments in appropriate infrastructure, all of which require political will as well as financial and human capital. The responses to the growing risks to coastal settlements brought on by climate change should include mitigation, migration and modification (McGranahan *et al.*, 2007).

Particularly as the need for action becomes more urgent, care will be needed to prevent government responses themselves from being inequitable or unnecessarily disruptive economically. Economically successful urbanisation is typically based on the decentralised decisions of economic enterprises and families, supported by their governments. When governments try to decide centrally where urban development should occur or where people should migrate, a range of political interests can intrude, favouring economically unviable locations and/or land use regulations that are particularly burdensome to the urban poor. Adaptation cannot be left to the market, but nor should it be left to arbitrary central planning (McGranahan *et al.*, 2007).

In many cases, there may be measures that can address present problems while also providing a means of adapting to climate change. These provide an obvious place to start, even if such coincidences of interest are unlikely to be sufficient to provide the basis for all of the adaptive measures needed.

Table 7.16 Population and land area in the low elevation coastal zone (LECZ) by national income, 2000

Income group	Population and land area in LECZ				Share of population and land area in LECZ			
	Population (million)		Land ('000 km²)		Population (%)		Land (%)	
	Total	Urban	Total	Urban	Total	Urban	Total	Urban
Low income	247	102	594	35	10	14	2	8
Lower middle income	227	127	735	70	11	14	2	8
Upper middle income	37	30	397	42	7	9	2	8
High income	107	93	916	129	12	12	3	9
World	618	352	2,642	276	10	13	2	8

Source: Adapted from McGranahan, Balk and Anderson, 2007.

At the national level, measures to support previously disfavoured inland urban settlements, away from the large cities on the coast, could reduce risks from climate change and also support a more balanced and equitable pattern of urban development. In China, for example, giving inland urban settlements the support needed to redress the imbalance caused by the creation of special economic zones along the coast would not only help reduce coastward migration but also reduce the increasingly severe regional inequalities that threaten China's national integrity (McGranahan *et al.*, 2007).

Alternatively, among coastal settlements in low income countries, those that find more equitable means to resolve the land problems that so often push their poorest urban residents to settle informally on unserviced and environmentally hazardous land (such as flood plains) will also be in a far better position to adapt to the risks of climate change. More generally, measures that support more efficient and equitable resolution of existing land issues are likely to provide a better basis for addressing the land issues brought on by climate change. Adaptive measures that respond to existing local needs, contribute to other development goals and can be locally driven are among the most likely to succeed (McGranahan *et al.*, 2007).

7.6 Management of urban carbon through spatial planning

In the context of cities, spatial planning is the key vehicle to deliver carbon management. The key principles affecting the ways in which spatial planning will be able to support innovation and the testing and acceptance of new or unfamiliar technologies will depend on the following:

1   *Leadership:* a clear statement of national policy objectives in terms of low carbon development which informs and validates all local planning objectives and which is capable of withstanding scrutiny upon challenge by developers.
2   *Recognition of the role of the development plan:* resource allocation and investment at the local level; and need for a national requirement that development plans set ambitious but achievable targets for low carbon energy production and use.
3   *Investment in knowledge:* strong, locally specific databases; an enhanced research programme and significant investment in measurement, monitoring and information sharing; and development of interactive communities of research and practice around the knowledge needed for low carbon development.
4   *Place based solutions:* bringing together energy and housing providers and developers to develop creative place-specific solutions, in close working relationships with community groups and local representatives.
5   *The importance of delivery vehicles:* local or otherwise appropriate development policy; and a national exemplars programme for zero carbon housing (Callcutt, 2007).
6   *Streamlining and integration of regulation:* planning permission needs to be clearly linked with environmental standards, certification and enforcement (e.g. building control standards and product certification); and ensuring that levels of control are proportionate and appropriate will involve clearly expressed permitted development for low impact technologies.

7.6.1 *Key challenges*

One of the key challenge to effective urban carbon management is the accurate estimation of emission at city scales. Parshall *et al.* (2010) listed four factors as key to successful urban scale emission inventories: consistency, spatial resolution, accounting framework, and attributes.

Consistency

The inventory should be built from systematic data collected for the entire country. Ideally, all raw data underlying the inventory should be derived from comparable energy sector data on location-specific fuel consumption. In practice, comparable sources of raw data for all sectors and fuels may be impossible to find, and some data may be derived from emissions models rather than from raw energy data. The consistency of the raw data within each sector is probably more important than the consistency across sectors. Data sources, and protocols for synthesising data into an inventory, should facilitate the release of inventories at regular intervals. Responsibility for data organisation and synthesis should be centralised at a single institution, preferably a government agency, to ensure that data products are recognised as authoritative and are available to the public.

Spatial resolution

The spatial resolution should match the smallest set of continuous administrative boundaries. Choosing continuous boundaries, rather than discontinuous boundaries, allows for complete coverage of the country at a high spatial resolution and allows analysis at multiple spatial scales.

Accounting framework

The inventory should be constructed according to a clearly defined accounting framework. The accounting framework should define the energy system perspective, including: whether the inventory will cover direct final consumption, total final consumption or total primary energy supply; how the inventory will allocate point source and non-point source data to localities; which fuels and/or sectors will be covered, and the scope of each sector; and how the inventory will partition data. For example, the inventory might categorise data by fuel, sector, sector and fuel, or sector and end use. Additional details on specific tools for these are discussed in Chapter 10.

Attributes

In addition to energy and CO_2 emissions data, the inventory should include consistent data for each locality on total population and spatial area. It also should designate the locality as urban or rural. These are the minimum attributes required to make meaningful cross-locality comparisons. Linking the inventory to additional climate, socio-demographic and economic indicators would help facilitate analysis of interactions between these factors and energy consumption and emissions.

7.7 Barriers to low carbon cities

Urban carbon management is in its early infancy. Most cities did not have comprehensive carbon management plans until about five years ago. Nevertheless, certain broad thematic barriers can be discerned.

7.7.1 *Management and culture*

These refer to the organisational ethos, habitual modes of practice, personalities and values present within municipal institutions, which may deeply influence the success of climate change action. Burch (2010) outlined the following management and cultural barriers to carbon management in cities:

- institutional funding structures and incentive programmes;
- the fit between the institutional arrangement and the problem it is intended to solve;
- the various levels of government claiming jurisdiction over a problem;
- codified rules and practices;
- antecedent development regulatory decisions.

These are further confounded by the municipal governance structures, levels of public awareness of climate change, and perception of the risk. In other words, contextual issues shape the environment within which the municipality functions and influence the values and priorities of the public (Burch, 2010).

7.7.2 *Planning and design barriers*

An appropriate urban form is a key factor to facilitate carbon management. Urban form, building uses, and their density patterns provide an over-arching system under which urban infrastructures related to buildings and transportation should be optimised. It is necessary to work within the urban form constraints in a comprehensive manner to avoid the 'rebound effect' where all positive gains could be offset by backsliding in other areas. Since changes to the urban form are slow, many of the design barriers are fixed.

7.7.3 *Data and technical barriers*

Dhakal and Shrestha (2010) suggested the following data and technical barriers to effective urban carbon management:

- data and information gaps;
- developing long term scenarios;
- establishing a consistent urban carbon accounting framework;
- understanding of the urban system dynamics;
- interaction of urban activities related to carbon emissions across the multiple system boundaries;
- formulating appropriate policies;
- operationalising the policy instruments.

However, a fully agreed framework and methods for such inventories of cities' greenhouse gas emissions are yet obscure, and this creates difficulties in comparing existing studies in cities. In order to arrive at any agreed framework, one of the key hurdles is the difficulties in setting the system boundary for an open system such as a city. Many studies and city action plans have used 'territorial' protocols (such as the IPCC or revised IPCC methods) to assign emissions to cities. However, approaches differ in terms of accounting for urban activities beyond territorial boundaries (for example, inter-city mobility by road, marine and air transport) (see Dhakal and Shrestha, 2010).

It is also necessary to undertake a comprehensive approach to allocate a city's carbon responsibility by including the upstream and downstream processes of connected socioeconomic systems and the indirect lifecycle related emissions. For example, the direct emissions account for only about 20 per cent of the overall upstream emissions necessary to sustain the input side of the economic production process in Singapore (Dhakal and Shrestha, 2010). Such a consumption oriented approach to urban carbon management is needed but not much pursued by the research community and also not addressed by the policy communities. A proper representation of a city's system boundary is essential for allocating appropriate responsibility for city carbon management, for policy making and for devising the effective carbon mitigation regimes.

Comparing cities for their carbon emissions, activities and policies needs a very detailed and careful look. A comparative perspective provides important insights but is often challenging, especially in case of cities where information is scarce and unconsolidated.

7.7.4 Individual lifestyles and behaviour change

Ultimately, much of the effort to reduce carbon at urban scales encounters strong barriers in the form of lifestyles and behaviours. Key problems include social dilemmas, social conventions, socio-technical infrastructures and the helplessness of individuals. In this light, recent literature suggests that more focus should be placed on the community level and that energy users should be engaged in the role of citizens and not only that of consumers (Heiskanen *et al.*, 2010). Low carbon communities could provide the supportive context for individual behavioural change. Heiskanen *et al.* (2010) suggested different communities for such a purpose: geographical communities as well as sector based, interest based and 'smart mob' communities.

References

All websites were last accessed on 30 November 2011.

Akbari, H. and Levinson, R. 2008. Evolution of cool-roof standards in the US. *Advances in Building Energy Research*, **2**, pp. 1–32.

Alexandri, E. and Jones, P. 2008. Temperature decreases in an urban canyon due to green walls and green roofs in diverse climates. *Building and Environment*, **43**, pp. 480–493.

Bartlett, S., Dodman, D., Hardoy, J., Satterthwaite, D. and Tacoli, C. 2011. Social aspects of climate change in urban areas in low- and middle-income nations. Contribution to the World Bank 5th Urban Research Symposium, Cities and Climate Change: Responding to an Urgent Agenda. Available at: www.dbsa.org/vulindlela/Papers%20Library/Session1_Satterthwaite.pdf.

BMA (Bangkok Metropolitan Authority). 2007. *Action Plan on Global Warming Mitigation 2007–2012*. Bangkok: BMA. Available at: http://www.baq2008.org/system/files/BMA+Plan.pdf.

Bowler, D.E., Buyung-Ali, L., Knight, T.M. and Pullin, A.S. 2010. Urban greening to cool towns and cities: a systematic review of the empirical evidence. *Landscape and Urban Planning*, **97**, pp. 147–155.

Brinkhoff, T. 2011. City population. Available at: www.citypopulation.de.

Brown, M.A., Southworth, F. and Sarzynski, A. 2008. *Shrinking the Carbon Footprint of Metropolitan America*. Washington, DC: Brookings Institution, Metropolitan Policy Program. Available at: http://www.brookings.edu/reports/2008/05_carbon_footprint_sarzynski.aspx.

Burch, S. 2010. In pursuit of resilient, low carbon communities: an examination of barriers to action in three Canadian cities. *Energy Policy*, **38**, pp. 7575–7585.

Callcutt, J. 2007. *The Callcutt Review of Housebuilding Delivery*. London: Department for Communities and Local Government. Available at: http://www.communities.gov.uk/documents/housing/pdf/callcutt-review.pdf.

Correa, C. 1989. *The New Landscape: Urbanization in the Third World*. London: Butterworth Architecture.

Crutzen, P. 2004. The growing urban heat and pollution 'island' effect: impact on chemistry and climate. *Atmospheric Environment*, **38**, pp. 3539–3540.

Dhakal, S. and Shrestha, R.M. 2010. Bridging the research gaps for carbon emissions and their management in cities. *Energy Policy*, **38**, pp. 4753–4755.

Dodman, D. 2009. Blaming cities for climate change? An analysis of urban greenhouse gas emissions inventories. *Environment and Urbanization*, **21**, pp. 185–201.

Emmanuel, R. 2005. *An Urban Approach to Climate Sensitive Design: Strategies for the Tropics*. London: E & FN Spon Press.

Emmanuel, R. and Fernando, H.J.S. 2007. Urban heat islands in humid and arid climates: role of urban form and thermal properties in Colombo, Sri Lanka and Phoenix, USA. *Climate Research*, **34**, pp. 241–251.

Emmanuel, R. and Johansson, E.J. 2006. Influence of urban morphology and sea breeze on hot humid microclimate: the case of Colombo, Sri Lanka. *Climate Research*, **30**, pp. 189–200.

Emmanuel, R., Johansson, E.J. and Rosenlund, H. 2007. Urban shading – a design option for the tropics? A study in Colombo, Sri Lanka. *International Journal of Climatology*, **27**, pp. 1995–2004.

Emmanuel, R. and Krüger, E. 2012. Urban heat island and its impact on climate change resilience in a shrinking city: the case of Glasgow, UK. *Building and Environment*, **53**, pp. 137–149.

Erell, E. 2008. The application of urban climate research in the design of cities. *Advances in Building Energy Research*, **2**, pp. 95–121.

Gehl, J. 1989. A changing street life in a changing society. *Places*, **6**, pp. 8–17.

Glaeser, E.L. and Kahn, M.E. 2010. The greenness of cities: carbon dioxide emissions and urban development. *Journal of Urban Economics*, **67**, pp. 404–418.

Gober, P., Brazel, A., Quay, R., Myint, S., Grossman-Clarke, S., Miller, A. and Rossi, S. 2010. Using watered landscapes to manipulate urban heat island effects: how much water will it take to cool Phoenix? *Journal of the American Planning Association*, **76**, pp. 109–121.

Golob, T.F. and Brownstone, D. 2005. *The Impact of Residential Density on Vehicle Usage and Energy Consumption*, Working Paper Series UCI-ITS-WP-05-1. Irvine: Institute of Transportation Studies, University of California. Available at: escholarship.org/uc/item/91v8x4qq.

Gromke, C., Buccolieri, R., Di Sabatino, S. and Ruck, B. 2008. Dispersion study in a street canyon with tree planting by means of wind tunnel and numerical investigations – evaluation of CFD data with experimental data. *Atmospheric Environment*, **42**, pp. 8640–8650.

Hardin, G. and Badden, J. 1977. *Managing the Commons*. San Francisco: W.H. Freeman and Co.

Havlick, S.W. 1983. An account of walking as a way of life in three European cities. *Ekistics*, **302**, pp. 351–355.

Heiskanen, E., Johnson, M., Robinson, S., Vadovics, E. and Saastamoinen, M. 2010. Low-carbon communities as a context for individual behavioural change. *Energy Policy*, **38**, pp. 7586–7595.

Hollander, J.B., Pallagst, K., Schwarz, T. and Popper, F.J. 2009. Planning shrinking cities. Available at: policy.rutgers.edu/faculty/popper/ShrinkingCities.pdf.

Holtzclaw, J. 2004. *A Vision of Energy Efficiency*. Washington, DC: American Council for an Energy-Efficient Economy. Available at: http://www.aceee.org/proceedings-paper/ss04/panel09/paper06.

Hoornweg, D., Sugar, L. and Gómez, C.L.T. 2011. Cities and greenhouse gas emissions: moving forward. *Environment and Urbanization*, **23**, pp. 207–227.

Hwang, R.-L., Lin, T.-P., Cheng, M.-J. and Lo, J.-H. 2010. Adaptive comfort model for tree-shaded outdoors in Taiwan. *Building and Environment*, **45**, pp. 1873–1879.

Hwang, R.-L., Lin, T.-P and Matzarakis, A. 2011. Seasonal effects of urban street shading on long-term outdoor thermal comfort. *Building and Environment*, **46**, pp. 863–870.

IEA (International Energy Agency). 2010. *World Energy Outlook 2010*. Paris: IEA. Available at: http://www.worldenergyoutlook.org/docs/weo2010/WEO2010_es_english.pdf.

Illich, I. 1982. *Gender*. New York: Pantheon Books.

Jabareen, Y.R. 2006. Sustainable urban forms: their typologies, models, and concepts. *Journal of Planning Education and Research*, **26**, pp. 38–52.

Jacobson, M.Z. 2009. Review of solutions to global warming, air pollution, and energy security. *Energy and Environmental Science*, **2**, pp. 148–173.

JaGBC (Japan Green Build Council) and JSBC (Japan Sustainable Building Consortium). 2006. *CASBEE for Heat Island (CASBEE-HI)*. Japan: JaGBC and JSBC. Available at: http://www.ibec.or.jp/CASBEE/index.htm.

Kalnay, E. and Cai, M. 2003. Impact of urbanization and land-use change on climate. *Nature*, **423**, pp. 528–531.

Kennedy, C., Steinberger, J., Gasson, B., Hansen, Y., Hillman, T., Havranek, M., Pataki, D., Phdungsilp,

A., Ramaswami, A. and Mendez, G.V. 2009. Greenhouse gas emissions from global cities. *Environmental Science and Technology*, **43**, pp. 7297–7302.

Lai, L.-W. and Cheng, W.-L. 2010. Air temperature change due to human activities in Taiwan for the past century. *International Journal of Climatology*, **30**, pp. 432–444.

Lal, D.M. and Pawar, S.D. 2011. Effect of urbanization on lightning over four metropolitan cities of India. *Atmospheric Environment*, **45**, pp. 191–196.

Lamptey, B. 2010. An analytical framework for estimating the urban effect on climate. *International Journal of Climatology*, **30**, pp. 72–88.

Landsberg, H.E. 1981. *The Urban Climate*. New York: Academic Press.

Lebel, L.A., Garden, P., Banaticla, M.R.N., Lasco, R.D., Contreras, A., Mitra, A.P., Sharma, C., Nguyen, H.T., Ooi, G.-L. and Sari, A. 2007. Integrating carbon management into the development strategies of urbanizing regions in Asia. *Journal of Industrial Ecology*, **11**, pp. 61–81.

Lu, X., Chow, K.-C., Yao, T., Lau, A.K.H. and Fung, J.C.H. 2010. Effects of urbanization on the land sea breeze circulation over the Pearl River Delta region in winter. *International Journal of Climatology*, **30**, pp. 1089–1104.

March, L. and Martin, L. 1972. *Urban Space and Structure*. Cambridge: Cambridge University Press.

Mayor of London (2007). *Action Today to Protect Tomorrow: The Mayor's Climate Change Action Plan*. London: Mayor of London. Available at: www.lowcvp.org.uk/assets/reports/London%20-%20climate%20change%20action%20plan.pdf.

McGranahan, G., Balk, D. and Anderson, B. 2007. The rising tide: assessing the risks of climate change and human settlements in low elevation coastal zones. *Environment and Urbanization*, **19**, pp. 17–37.

MOEWR (Ministry of Environment and Water Resources). 2007. *Singapore's National Climate Change Strategy*. Singapore: MOEWR. Available at: http://app.mewr.gov.sg/data/ImgUpd/NCCS_Full_Version.pdf.

Ng, E. 2009. Policies and technical guidelines for urban planning of high-density cities – air ventilation assessment (AVA) of Hong Kong. *Building and Environment*, **44**, pp. 1478–1488.

Nieuwolt, S. 1966. The urban microclimate of Singapore. *Journal of Tropical Geography*, **22**, pp. 30–37.

Oswalt, B.P. and Rieniets, T. 2007. Shrinking cities: global study. Available at: http://www.shrinkingcities.com/globaler_kontext.0.html?&L=1.

Parker, D.E. 2010. Urban heat island effects on estimates of observed climate change. *WIREs Climate Change*, **1**, pp. 123–133.

Parshall, L., Gurney, K., Hammer, S.A., Mendoza, D., Zhou, Y. and Geethakumar, S. 2010. Modeling energy consumption and CO_2 emissions at the urban scale: methodological challenges and insights from the United States. *Energy Policy*, **38**, pp. 4765–4782.

PlaNYC. 2011. *PlaNYC Update April 2011: A Greener, Greater New York*. New York: Mayor's Office. Available at: http://www.nyc.gov/html/planyc2030/html/publications/publications.shtml.

Priyadarsini, R., Wong, N.H. and Cheong, K.W.D. 2008. Microclimatic modelling of the urban thermal environment of Singapore to mitigate urban heat island. *Solar Energy*, **82**, pp. 727–745.

Rieniets, T. 2009. Shrinking cities: causes and effects of urban population losses in the twentieth century. *Nature and Culture*, **4**, pp. 231–254.

Salim, S.M., Cheah, S.C. and Chan, A. 2011. Numerical simulation of dispersion in urban street canyons with avenue-like tree plantings: comparison between RANS and LES. *Building and Environment*, **46**, pp. 1735–1746.

Singapore Government. 2011. Vehicle ownership in Singapore. Available at: http://app.mot.gov.sg/Land_Transport/Managing_Road_Use/Vehicle_Ownership.aspx.

Sovacool, B.K. and Brown, M.A. 2010. Twelve metropolitan carbon footprints: a preliminary comparative global assessment. *Energy Policy*, **38**, pp. 4856–4869.

Spangenberg, J., Shinzato, P., Johansson, E. and Duarte, D. 2009. Simulation of the influence of vegetation on microclimate and thermal comfort in the city of São Paulo, Brazil. *Rev. SBAU, Piracicaba*, **3**, pp. 1–19.

Synnefa, A., Karlessi, T., Gaitani, N., Santamouris, M., Assimakopoulos, D.N. and Papakatsikas, C. 2011. Experimental testing of cool colored thin layer asphalt and estimation of its potential to improve the urban microclimate. *Building and Environment*, **46**, pp. 38–44.

Tay Kheng Soon. 2001. Rethinking the city in the tropics: the tropical city concept. In *Tropical Architecture: Critical Regionalism in the Age of Globalization*, ed. A. Tzonis, B. Stagno and L. Lefaivre, pp. 266–306. Chichester: Wiley Academic.

Taylor, B. and Guthrie, P. 2008. The first line of defence: passive design at an urban scale. *Proceedings of Conference: Air Conditioning and the Low Carbon Cooling Challenge, Cumberland Lodge, Windsor, UK, 27–29 July 2008*. London: Network for Comfort and Energy Use in Buildings. Available at: http://nceub.org.uk.

Tayyaran, M.R. and Khan, A.M. 2003. The effects of telecommuting and intelligent transportation systems on urban development. *Journal of Urban Technology*, **10**, pp. 87–100.

TMG (Tokyo Metropolitan Government). 2007. *Tokyo Climate Change Strategy: A Basic Policy for the 10-Year Project for a Carbon-Minus Tokyo*. Tokyo: TMG. Available at: http://www.kankyo.metro.tokyo.jp/climate/attachement/tokyo-climate-change-strategy_2007.6.1.pdf.

UNPD (United Nations, Department of Economic and Social Affairs, Population Division). 2010. *World Urbanization Prospects: The 2009 Revision*, CD-ROM edn – data in digital form, POP/ DB/WUP/Rev. 2009.

Wright, C.A. 1992. *Fast Wheels, Slow Traffic: Urban Transport Choices*. Philadelphia, PA: Temple University Press.

Yamamoto, Y. 2006. Measures to mitigate urban heat islands. *Science and Technology Trends: Quarterly Review*, **18**, pp. 65–83. Available at: http://www.nistep.go.jp/achiev/ftx/eng/stfc/stt018e/qr18pdf/STTqr1806.pdf.

Yang, F., Lau, S.S.Y. and Qian, F. 2010. Summertime heat island intensities in three high-rise housing quarters in inner-city Shanghai, China: building layout, density and greenery. *Building and Environment*, **45**, pp. 115–134.

Yang, L. and Li, Y. 2009. City ventilation of Hong Kong at no-wind conditions. *Atmospheric Environment*, **43**, pp. 3111–3121.

Zhao, C., Fu, G., Liu, X. and Fu, F. 2011. Urban planning indicators, morphology and climate indicators: a case study for a north–south transect of Beijing, China. *Building and Environment*, **46**, pp. 1174–1183.

8   Operational and embodied carbon in buildings

Measures to reduce carbon in the built environment take the form of the following: reduce operational energy (and therefore carbon) in buildings; reduce embodied carbon in building materials and construction; and switch to low carbon fuels or use renewable energy (Levine *et al.*, 2007). Chapter 4 outlined key approaches and technologies to low carbon or renewable fuels and micro-generation at site. Chapters 5–7 dealt largely with ways to reduce the need for operational energy in buildings. In this chapter we explore carbon embodied in buildings as well as broad strategies to manage energy in use.

In the year a project is built or installed, 13–18 per cent of the building's total carbon footprint is released, while the remainder of the carbon footprint is the operational carbon released over the life of the project (UNEP, 2007). While the embodied carbon appears to be a small part of the overall carbon problem in buildings, there are specific instances where embodied carbon could assume greater importance. Embodied carbon is becoming important in low/zero energy buildings, where operational energy requirements are greatly reduced. For example, Thormark (2002) found the energy embodied in a low energy Swedish single-family house to be nearly 40 per cent of the whole life (50 years) energy requirements. Similarly, the proportion of embodied energy in buildings in developing countries could be very high, considering the very low operational energy use (Levine *et al.*, 2007). A single-minded focus on operational carbon may not be enough in the context of this book (focusing on LZC buildings) as well as in the international context.

8.1 Embodied carbon

Embodied carbon is defined as the carbon cost (in CO_2 or CO_2e) of construction or manufacturing. It refers to the total primary energy consumed (and carbon dioxide released) from

Table 8.1 Operational carbon emission in a typical UK commercial building

Building function	Fraction of total carbon emission (%)
Lighting	19
Mechanical ventilation	16
Office equipment	16
Comfort cooling	10
Infiltration	9
Heat loss via windows	6
Pumps and controls	5
Other equipment	5
Catering	4
Losses through:	
Flue	2
Wall	2
Roof	2
Floor	2
Hot water	2

Source: Based on RAE, 2010.

direct and indirect processes associated with products or services, including material extraction, manufacture, transportation and any fabrication before the product is ready to leave the factory gate. Commonly the cost is calculated for cradle to gate (material extraction to factory gate), or sometimes for cradle to site (material extraction to construction site), but a more representative figure would be from calculating the whole life cost (cradle to grave). However, this is problematic because of the uncertainty in knowing how a building will be demolished (and the volume of materials recovered and recycled) at the end of its life.

Owing to the lack of a formal definition, the terms 'embodied carbon' and 'embedded carbon' are often used interchangeably; therefore when quoting a figure for either of these terms it is critically important to understand the boundaries for which it has been calculated. Particularly with respect to the construction industry there is an advantage in retaining both terms, albeit with subtly different definitions. Therefore, based on Ainger *et al.* (2008), for the purposes of this textbook we have adopted the following definitions: embodied carbon refers to the whole life carbon cost of construction or manufacture – cradle to grave where possible, and cradle to gate or site otherwise, with clarification; and embedded carbon refers to the carbon cost of maintaining a building or building component over its operational lifespan, for example the carbon cost of replacing a pump in a water pumping station over the operational lifetime of the station. The distinction is useful, as when choosing the lowest carbon option for a project it may be necessary to decide between one option that has a high upfront carbon cost for construction but lower downstream carbon costs for maintenance, and another with a lower upfront cost but higher downstream costs for maintenance. An example of this might be choosing between designing a passively ventilated building (with a high upfront carbon cost due to higher thermal mass) and a 'climate responsive' light build design incorporating mechanical ventilation.

In terms of whole life carbon emissions, Yohanis and Norton (2002) identified four main components:

- *initial embodied energy:* the energy required to produce the building initially, which includes the energy used for the abstraction, the processing and the manufacture of the materials of the building as well as their transportation and assembly on site;

- *recurring embodied energy:* the energy needed to refurbish and maintain the building over its lifetime;
- *operational energy:* the energy used to operate the building, in other terms to provide heating, cooling and lighting, and power the various appliances of the building;
- *demolition energy:* the energy to demolish and dispose of the building at the end of its life.

8.1.1 Carbon in building materials

Currently the most comprehensive source of figures for carbon embodied in building materials is Bath University's Inventory of Carbon and Energy (ICE) (Hammond and Jones, 2011). Until recently ICE was free to download and supported by a wiki, but at the time of writing these services had been withdrawn. However, the inventory is still available online (Hammond and Jones, 2011), and new updates are expected. Table 8.2 gives general figures for embodied energy and carbon in common building materials, and the chapter uses these examples to highlight issues that need to be considered when selecting factors for calculating more specific figures. In addition, when calculating the carbon and energy embodied in building materials it is important to recognise that the fuel mixes used for generating heat and electricity and powering plant machinery will have a significant impact on the results. For imported materials and products it is important (where possible) to factor in the energy mix for the country of production, rather than use figures for the recipient nation as a proxy (and if these are used it should be clearly noted in all reporting). Finally, it should also be noted that when calculating the energy embodied in a construction project it is necessary to account for the additional carbon and energy costs attributable to transporting material or products to site, those attributable to the construction phase itself, and those attributable to the transportation, treatment and disposal or recycling of construction waste.

Concrete

After the burning of fossil fuels, the production of cement is the next largest global source of carbon emissions. However, the volume of cement in any given concrete mix or product is just

Table 8.2 Embodied carbon and energy in common building materials

Building material	Embodied carbon (KG CO_2/tonne)
Sandstone	64
Stone (average)	79
Granite	93
Marble	112
General concrete	100–130
General clay bricks	220
Slate	232
Timber	450–750
Facing bricks	520
General building cement	830
Glass (primary)	910
Steel: bar and rod	1,710
Steel: world average	1,950
Steel: galvanised sheet	2,820
Copper (tube and sheet)	2,710

Source: Derived from Ainger *et al.*, 2008; Crishna *et al.*, 2010; Hammond and Jones, 2011.

one of many factors that will determine its embodied carbon. Owing to the wide variation in the figures for different types of concrete, when calculating the carbon embodied in concrete it is critical to know exactly what type(s) will be used in order to produce accurate results. This is made easier because of the prevalence of concrete as a building material and the comparative wealth of figures available for different mixes and products, but the variety of concrete mixes (for example, the full or partial substitution of sand or aggregate with alternatives) can make producing a definitive result a problematic task for all but the most standard mixes or products.

Steel

Steel is another building material for which many figures exist for embodied carbon, according to factors such as type, recycled content, and final product (e.g. sheets or bars), but in the latter case the ICE's figures do not account for the cutting of steel products to size prior to or after leaving a steelworks. Calculating a definitive figure for the carbon embodied in any specific steel product is problematic because of several factors: the production and treatment or use of by-products from steel manufacturing; the production and use of excess energy generated on site at the steelworks; and the release of fugitive emissions from manufacturing, such as from the calcination of lime.

Timber

Timber is by far the most problematic material for carbon accounting. Aside from the many different varieties and products available, trees absorb different amounts of carbon at different stages in their lifecycle and under different environmental conditions. The justification of any particular set of boundary conditions is also highly problematic without knowing the source of the wood and how the material will be treated at the end of its life. In the former case it might be considered justifiable to use a lower figure for embodied carbon if the timber is from a sustainably managed forestry scheme, but this also ignores the wider issue of the environmental and ecological impacts of monoculture plantations. In the latter case the embodied carbon may vary significantly according to whether the material is burnt, sent to landfill or otherwise dealt with at the end of its life, and for this reason alone calculations should not deduct the carbon sequestered by the growth of the timber from any final results.

Stone

Compared to other building materials, calculating the carbon embodied in any particular type of stone or stone product is a comparatively simple task, particularly in countries where stone extraction and processing are done largely by machinery rather than manual labour. Because stone is a raw material that can be used straight out of the ground, and because primary processing (e.g. cutting into blocks) is often done on or near the extraction site, the basic figures for the carbon embodied in stone are small in comparison to that in other materials. Waste material is also commonly left on site for later use in back filling, although calculations should account for any use of chemicals to treat the material, and their removal from any waste prior to disposal. This means that, for any given use of stone, carbon embodied in it may be largely dictated by the mode and distance of transportation from gate to site (see also Crishna *et al.*, 2010).

Glass

For many applications it is also comparatively easy to calculate the carbon embodied in glass. The main uncertainties come from the release of fugitive carbon dioxide emissions when soda ash, lime and dolomite are melted during manufacture. However, it remains difficult to calculate accurate figures for more specialist products, such as toughened and polarising glass, which undergo additional treatments during the manufacturing process.

Copper

As the extraction of copper from virgin ore is a highly energy intensive process, the main variable in determining the carbon embodied in copper products is percentage of recycled material used in any given product. This is further complicated by the source of the recycled material and the quality of the final product, as recycling copper from lower grade sources and/or into higher grade products increases the energy used for processing. Today virtually all copper contains some recycled material, which varies annually owing to factors such as supply and demand.

Aluminium

Aluminium is the most abundant metal in the earth's crust, but also one of the most widely recycled. The reasons for this may be that it takes 95 per cent less energy to recycle aluminium than it does to extract it from virgin ore, and recycling it does not require industrial scale electrolysis (*Economist*, 2007). Aluminium has also become an iconic metal for recycling after being the focus of numerous high profile recycling campaigns.

Iron

Accounting for the carbon embodied in iron remains problematic because of its broad range of uses and lack of authoritative sources, with the added complication that some of the data collected to date may have included figures for material that was actually steel.

Plastics

Figures are available for the embodied energy in all the most commonly used forms of plastic, and many that are used widely but in small quantities. These figures vary notably between plastics with very similar names, and for more accurate accounting these need to be verified as consistent throughout the product chain.

Summary

Calculating embodied energy and carbon is a relatively new field in construction management, but such figures are increasingly used in decision making, for example to justify the use of locally sourced or more costly but lower carbon materials. Given sufficient knowledge and understanding of the material and its use (source, fuel mix, product type, transportation and so on) it is fairly simple to select appropriate figures to calculate justifiable results, albeit with some approximations and varying degrees of uncertainty. As for all other areas of carbon management it is critical to ensure that any calculations of embodied energy are defensible by justifying and reporting all boundaries, factors, approximations and assumptions used, and the levels of uncertainty involved. Finally, as the number and accuracy of conversion factors increase, developers of tools for calculating embodied energy and carbon are well advised to ensure the tools are designed to allow these factors to be updated frequently and easily, and ensure this is done periodically and using the most authoritative sources.

8.2 Energy management in buildings

8.2.1 Smart meters

The term 'smart meter' covers an increasingly wide range of devices used to inform the occupants of a building about their energy consumption and regularly transmit the data to utility companies. Smart meters are also used for monitoring gas and water consumption, and more advanced devices can be used to control building services and appliances remotely (see 8.2.2 below). The term is also frequently used to include those devices that display consumption data but do not broadcast it beyond the building, which may be justifiable for those meters capable of measuring or disaggregating the energy consumption attributable to individual building services or appliances. For more basic devices that do not transmit data beyond the home the term 'energy consumption indicator' (ECI) is used in literature and is perhaps more appropriate. For simplicity the term 'smart meter' is used here for both.

Smart meters have two main uses: to provide actual consumption data to utility companies to allow them to issue accurate bills (rather than estimated bills, which are frequently contested); and to inform occupants about how much energy they are using and what is responsible for this. Providing real-time consumption data to occupants has been found to be effective in reducing

Table 8.3 International status of smart meters

	Principal motives	*Smart metering status*	*Regulation*	*Comments*
California	Load management; peak reduction	In progress for electricity; gas 'piggybacking'	Large local monopolies; vertical integration	Some success with peak reduction; a low but increasing interest in demand reduction; some strong customer resistance
Italy	Fraud reduction; contractual power control and load limiting	Roll-out almost complete (electricity)	Slight competition, with ENEL dominant	Payback time of less than five years is claimed; no customer displays as yet
Malta	Fraud reduction; water conservation	Roll-out to begin soon	Monopoly	Demand reduction low down on the list of priorities; no customer displays
Netherlands	Demand reduction; load management	Mandatory roll-out halted; terms being renegotiated	Liberalized; networks own meters	Legal challenge on data privacy halted the roll-out; customer displays being developed as part of the offer
Ontario, Canada	Load management	Roll-out complete; time-of-use pricing now under way	Many local monopolies	Some successes with demand reduction from trials with in-home displays, but they are not rolled out with the smart meters
Sweden	Accurate billing	Roll-out complete	Liberalized; networks own meters	Some web-based feedback to customers; very few displays

Source: Darby, 2010.

energy consumption when used either in place of or in conjunction with other behavioural levers (Faruqui, Sergici and Sharif, 2010; Wood and Newborough, 2003). Basic smart meters usually display the amount of energy being consumed, the cost (financially and in CO_2), and how this compares to previous consumption, for example against that of the previous day or a monthly average. These meters rely on occupants learning about their energy consumption by switching services or appliances on and off and noting the changes, but more advanced devices (e.g. those that can be linked to home computers) can provide a greater range of data outputs. For more on human behaviour see Section 3.8.

Support for the wider roll-out of smart meters is based on the plethora of evidence that shows that enabling building occupants to have a greater understanding of and, critically, control over their energy (and water) consumption leads to demand reduction and/or optimisation of use (e.g. Ueno *et al.*, 2005; Willis *et al.*, 2010). Most importantly, as human behaviour is partly a construct of cultural and societal norms and values, these findings are borne out internationally, meaning that the influence is effective (to a greater or lesser extent) regardless of social and cultural differences.

Many countries around the world are now engaged in programmes to facilitate the mass roll-out of smart meters, including Sweden, Italy, the USA, Canada and Australia (see Table 8.3 for details). In the EU the Energy End Use Efficiency and Energy Services Directive (2006/32/EC) (also known as the Energy Services Directive) requires the installation of basic (real-time display only) meters in all new buildings and when existing meters are replaced. Other supporting policies to enable this roll-out include the UK's requirement that utility companies provide customers with basic meters free upon request (SDC, 2007). Although market penetration is still relatively low, and studies of the impact of basic meters have not found drastic reductions in consumption, e.g. around 7 per cent in the USA (Faruqui *et al.*, 2010), improvements in design and functionality of the devices show the potential for more significant savings, especially when combined with other technological measures. Meyers, Williams and Matthews (2010) highlight four measures that could be combined with smart meters to reduce energy consumption: programmable thermostats; zoned heating and cooling; remote control of HVAC systems; and outlet level appliance monitors that can automatically disconnect appliances to eliminate leakage currents. Whilst the current prevalence of these varies by country, they are all mature technologies that could potentially converge with each other and smart meters, and whilst the cost of a combined package for domestic use may be prohibitive at present this is expected to fall as demand increases. Further ahead, consumer demand could also be increased by designing systems for even greater levels of remote control, for example by using home wireless networks linked to mobile phones to allow HVAC systems and a wider range of appliances to be switched on and off, possibly even passively in response to locational data from the handset. However, even at present the limiting factors for smart meters are more human than technological: designing displays to be more effective in influencing behaviour, and selecting combinations of functions that are the most useful for consumers.

The psychology of smart meters

Despite the drive to roll out domestic smart meters in many countries, there remains considerable debate around how best to design these to enable householders to reduce consumption, and current evidence suggests there is no one-size-fits-all option. There is a wealth of evidence on how to design smart meters and, whilst this is often contradictory and sometimes counter-intuitive, the debate is centred on how, how much and how frequently data should be supplied to users in order to best achieve the desired effect in an average household. For example, Wood and Newborough (2003) considered the problem of displaying consumption in different units (kW, kWh, CO_2 and £) and whether to display these numerically and/or graphically. A prior assertion was that householders would be most likely to understand (and therefore act on) information displayed in £, but it was noted that, if occupants were using energy unnecessarily but the marginal cost of this was only a few pence, then this was unlikely to be a sufficient motivator to change that behaviour. Therefore displaying different measures of consumption that can show greater 'differences' may help counter this perception, even if the units are less familiar to users. Another option employed is to use data presented as histograms in addition to or in place of numeric displays, in order to divert users from focusing on changes that appear negligible when expressed numerically.

Another key problem is avoiding information overload, whilst at the same time ensuring enough is provided to enable users to understand and change their energy consumption habits. One stumbling block for studies aimed at informing the design of smart meters is that there may be an inherent bias towards recruiting volunteers who are 'early adopters' of technology, who may be more comfortable with greater volumes of information and/or more aware of their energy consumption. There is also the problem that users read the displays in different ways depending on the information being sought (potentially leading to different outcomes), and it is difficult to design a display to meet these needs. Wood and Newborough (2003, 2007) found that users may be seeking a specific quantitative reading (e.g. average daily consumption), to observe a rate of change over time, or to observe the change when a specific appliance is switched on or off. However, as mentioned above, some of these can result in unintended effects that may be contrary to the intended use of the meters. The work of Hargreaves, Nye and Burgess (2010) gives a fascinating insight into the psychology of human interaction with smart meters (both basic and more advanced models). Most importantly, the study found that, even amongst self-professed early adopters, responding to the information and seeing their consumption fall did little to counter the perception that these small changes were anything more than 'tinkering around the edges'. However, it is the very human stories that emerged from their use that shed light on the potential benefits (or otherwise) of rolling out smart meters to all homes. An important objection that has been raised against smart meters is their potential role in a 'surveillance state' in which increasing amounts of personal data are collected by and shared between government agencies and the private sector. Some studies (e.g. Darby, 2005) have found that introducing an element of competition between households may serve to reduce consumption further, but these have tended to be amongst groups of energy aware participants who are more open to disclosing this information. Nevertheless, some housing developers are already incorporating the idea into their designs (e.g. PortZED in Shoreham, UK). However, Hargreaves *et al.*'s findings should be treated as a word of caution. Even though the displays were kept within households and the readings were not disclosed, they were still frequently the cause of arguments – in one case because a husband admitted to using the device to remotely monitor when his wife was cooking dinner!

8.2.2 Intelligent energy management

Intelligent energy management systems (IEMs) are the non-domestic big brother of smart meters. At a very basic level they include 'dumb' systems, such as motion sensors to control lighting, and 'closed loop' monitoring systems that simply display aggregated consumption information to building managers. At the more intelligent end of the spectrum they are becoming increasingly sophisticated, for example measuring energy consumption of individual building services, controlling these remotely via networks linking geographically disparate buildings, and also incorporating the monitoring and control of renewable energy, CHP and energy storage systems to optimise the matching of supply and demand. A common example of an IEM is a street lighting network that is controlled by timers and/or light sensors, and uses PV cells and local storage to generate electricity during the day for use during the night. At present most IEMs rely on some degree of human input, but as artificial intelligence improves this will further expand functionality and decrease the need for human input.

Although IEMs are still in their relative infancy, studies are already modelling their impact on reducing energy consumption. One such study by Papagiannis *et al.* (2007) found that for the EU-15 nations, and assuming a logical rate of market penetration, even a fairly basic and cheap IEM could achieve a reduction of 1–4 per cent in primary energy, a reduction of 1.5–5 per cent in CO_2 emissions, and a 2–8 per cent saving in investment costs for power generation expansion. The study was also able to justify the assertion that such innovative systems may be attractive to end users and aid the implementation of global energy saving policies. However, the findings are not consistent across all sectors – potential savings were highest for industry, but this was hampered by the likely low market penetration rate, with the greatest overall potential being for street lighting and the 'tertiary' sector (essentially public and agricultural buildings and services).

As the use of IEMs gains traction it is important to note its benefits for supporting distributed micro-generation and CHP. Studies into building energy management have traditionally focused on energy conservation, but the contribution of micro-generation to meeting energy demands is increasing with the drive towards 'energy positive' buildings, necessitating the use of more sophisticated IEMs to manage supply and demand, and capable of managing the export of excess energy to local networks.

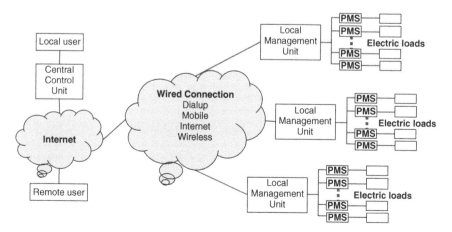

Figure 8.1 Basic layout of an energy consumption management system

8.2.3 The role of behaviour and attitudes

The importance of delivering effective behaviour change in reducing emissions is widely recognised, as is the need to deliver it across all sectors and sections of society (EST, 2008; Scottish Government, 2009). This also demonstrates the need for greater understanding, and a strategy that goes significantly further and is more nuanced than the policy initiatives that have sought to target specific public behavioural changes in the past. Such a strategy should also take into account the widely different attitudes towards climate change, and therefore enabling effective behaviour change also requires understanding and addressing these differences. It must be recognised that even those most predisposed to adopting pro-environmental behaviours can behave far from rationally – or conversely it may be completely rational not to adopt a 'pro-environmental' behaviour if the costs and implications of doing so are deemed to be unacceptable, for example for financial reasons or any additional time required.

Highly successful behaviour change initiatives include using seatbelts (Jochelson, 2007) and tackling drink driving (Mann *et al.*, 2001), speeding (Pilkington and Kinra, 2005) and smoking (Adshead and Thorpe, 2007). However, these are single-behaviour changes achieved by awareness raising backed by legislation and new laws, and even for these it is useful to note the length of time each took to take hold.

The energy efficiency of a household is the product of many behaviours, as well as the wider demographic and socioeconomic context and composition of the household. This complexity means that studies into energy consumption behaviour tend to be split into those that focus on how behaviour influences consumption and those that study the effectiveness of specific behaviour change interventions (for example, the plethora of work around designing smart meters).

A key problem for policy makers is that whilst significant gains could be made from targeting known (modellable) behavioural influences on consumption it is generally neither possible nor desirable to influence these behaviours directly. Conversely, scaling up the results from many studies of behavioural interventions risks overestimating potential savings for a number of reasons, for example participant self-selection bias, the Hawthorne effect (the impact on behaviour of the knowledge of being studied), and the replicability of a particular intervention at a much larger scale.

Although research has found that behaviour change can achieve energy savings of 5–15 per cent (Martiskainen, 2007), it has also found very high levels of variability in the influence of behaviour on heating and electricity and water consumption (Gill *et al.*, 2010). This underlying complexity and the (current) lack of large datasets make it unwise to scale up the results of specific behaviour change interventions to wider changes in total household energy efficiency – and therefore emissions savings. However, there is a wealth of evidence about what works from wider studies on behaviour change – both on specific measures and how to implement them most effectively – that can be employed for reducing emissions from households.

Our reading of the numerous common themes that run through the evidence base for the energy use effects of behaviour is as follows:

- Facilitating conditions for behaviour change is at least as important as trying to influence it directly.
- Behaviour change is most effective when a number of levers are pulled in a coherent, coordinated and systematic way.
- Government and its wider partners need to be seen to be leading by example.
- Behaviour change needs to be coordinated across sectors and sections of society.
- Changing social norms is key, whether this is through provision of infrastructure and services or through regulation.

- Interventions are most effective when tailored to the local or community level, or to specific groups of people.
- Targeting multiple contexts, moments of lifestyle transition and institutional or infrastructural pressure points is effective.
- Policies aimed at changing behaviour need to be simple and transparent.
- Climate change is generally not a useful motivator, and for some will have the opposite effect.
- Positive images of pro-environmental behaviour to promote non-environmental messages are effective, for example promoting the health benefits of walking and cycling.
- Awareness raising and information provision alone does not work effectively, but does work better when tailored to the target audience.
- Knowing and responding to the target audience are essential for success.
- The messenger must be trusted by the target audience.
- More needs to be done to develop behaviour change initiatives in parallel with technological change in order to maximise the benefits of both.
- The effectiveness of financial interventions partly depends on the ability of people to gauge their future circumstances, and this is particularly difficult when their current financial situation is unpredictable.
- Feedback, for example from home energy surveys or the displays on smart meters, needs to be designed to prompt action. For instance, most households will be more receptive to financial savings than to energy or emissions savings.
- Greater cumulative outcomes can be achieved by targeting mutually reinforcing behaviours.
- Ideally, a substitute behaviour needs to be more attractive than the default.

Whilst this list is extensive and the number of schemes and projects aimed at changing energy consumption behaviour is now growing, the field is still relatively new and poorly understood. This presents significant opportunities for research, innovation and creativity that should yield further emissions reductions.

Another challenge here is how to reinforce such behaviours so they 'stick' over the long term and become habits rather than conscious actions, for example remembering to switch off unused lights and appliances (Jackson, 2005). The evidence for how long any given intervention 'sticks' is limited and contested, especially in the case of reported behaviour. Therefore it may be more useful to prioritise single-behaviour changes that do not require regular repetition but which may reinforce other behaviours, such as the purchasing of energy efficient appliances, whilst progressing with the development of effective attitudinal interventions.

It is one problem to understand human behaviour as regards energy use and yet another to influence it effectively. Whilst a full discussion of this subject is beyond the scope of this book, there are some simple guidelines that should be taken into account when attempting to influence behaviour to reduce energy consumption. First of all, it is important to recognise that energy inefficiency is not a single behaviour that can be addressed in the same way as negative behaviours such as smoking or drink driving. This helps explain why information programmes on energy reduction are rarely effective (Dulleck and Kaufmann, 2004), and it is worth noting that, in Wood and Newborough's (2003) study of smart meters, those who were supplied with both a meter and an information pack tended to attribute their savings to the meters rather than the packs.

Switch what off?

Figure 8.2 shows a bank of light switches in an open plan office in the UK. These are situated close to desks and are labelled with 'Switch it off' stickers from the UK government's 'Act on CO_2' information campaign. They control fluorescent lights, so dimmer switches cannot be installed, but in all other ways they are an example of good practice in energy efficiency – bar one. Can you spot it?

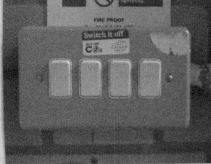

Figure 8.2 Light switch boxes – which to switch off?

Photo credit: Denis Fan, IESD

How do you know which switch controls which light? Remember that this is an open plan office. Users of the building reported that not wanting accidentally to switch *someone else's* light on or off discouraged them from using the switches. Simply labelling the switches would solve this problem – which goes to show that human behaviour is complex and open to subtle influences, but also that listening to feedback from building occupants can reveal simple ways of reducing energy consumption.

8.3 Key issues faced by the construction industry

Apart from the difficulties faced by designers and users in reducing the operational and embodied carbon in buildings, there are construction industry-specific barriers and difficulties that need to be overcome to quantify the nature and scale of the carbon-in-use problem better.

8.3.1 Lifecycle carbon versus energy reduction

The main tool for making lifecycle carbon savings attractive to the building owners and occupiers is to 'tax' carbon emissions. On the other hand, countries and jurisdictions utilising highly polluting processes for electricity generation (for example, coal) will benefit the most by focusing their energies on building energy efficiency improvements. Others may not see such improvements. The trade-off between cleaning up energy production and building efficiency improvement is a fine balancing act (Kneifel, 2010) (see also Figure 8.3 for an illustration of the problem).

It is also necessary to consider the correct time horizons to evaluate the lifecycle carbon reductions. Some approaches to low carbon design and construction are cost-effective regardless of time horizons; many are not. The realised costs of a building are overlooked when the future costs of operating and maintaining the building are not taken into account. Kneifel (2010) points

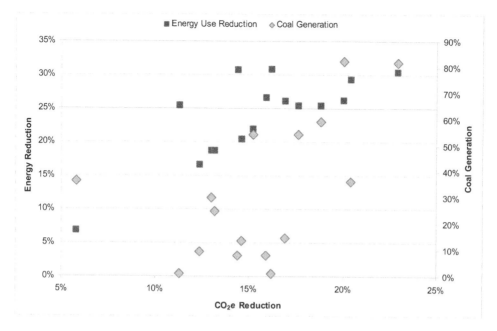

Figure 8.3 Energy and emission reduction in the United States

Source: Based on data from Kneifel, 2010.

out that more building type–location combinations find it cost-effective to adopt the most energy efficient building design alternative, with the greatest change occurring between the 1- to 10-year and 10- to 25-year study periods.

8.3.2 Barriers to reducing operational carbon in the housing sector

Apart from the numerous design barriers to local carbon buildings enumerated in Chapters 5–7, there are specific barriers relating to the construction industry in the operation and maintenance aspects of carbon in buildings. Among other things, Osmani and O'Reilly (2009) found the following pertaining to the construction industry in the UK:

1   There is a belief that 'unreliable' technologies are being installed to the detriment of profit.
2   Volume house builders in the UK tend to use a range of standard house sets across their developments to help reduce costs and defects and, as a consequence, they are reluctant to adopt policies which require excessive design changes.
3   An unwillingness to implement untested or new sustainable materials and products is an acute problem in the housing sector, where traditional attitudes restrict the uptake of innovations.
4   There is a widespread perception that there is currently a lack of demand for sustainable properties amongst the general public.
5   It is perceived that there are increased costs in achieving the high building standards associated with low and zero carbon homes (Osmani and O'Reilly, 2009).

Energy in use could be reduced only if firm but balanced legislation were in place. Reflecting this, Banfill and Peacock (2007) called for the UK Government to move away from energy conservation policy making towards the production of detailed yet balanced legislation to help create a market for sustainability and thus reduced carbon in buildings both in their design and in the operational and embodied phases.

References

All websites were last accessed on 30 November 2011.

Adshead, F. and Thorpe, A. 2007. The role of the government in public health: a national perspective. *Public Health*, **121**, pp. 835–839.

Ainger, C., Baker, K., Crishna, N., Cziganyik, N., Fenn, T., Johnson, A., Jowitt, P., Robertson, C. and Smith, C. 2008. *Carbon Accounting in the UK Water Industry: Guidelines for Dealing with 'Embodied Carbon' and Whole Life Carbon Accounting*. London: UKWIR.

Banfill, P.F.G. and Peacock, A.D. 2007. Energy efficient new housing – the UK reaches for sustainability. *Building Research and Information*, **35**, pp. 426–436.

Crishna, N., Goodsir, S., Banfill, P. and Baker, K. 2010. *Embodied Carbon in Natural Building Stone in Scotland*. Edinburgh: Historic Scotland.

Darby, S. 2005. Social learning and public policy: lessons from an energy-conscious village. *Energy Policy*, **34**, pp. 2929–2940.

Darby, S. 2010. Smart metering: what potential for householder engagement? *Building Research and Information*, **38**, pp. 442–457.

Dulleck, U. and Kaufmann, S. 2004. Do customer information programs reduce household electricity demand? The Irish program. *Energy Policy*, **32**, pp. 1025–1032.

Economist. 2007. Case history: the truth about recycling, 7 June. Available at: http://www.economist.com/node/9249262.

EST (Energy Saving Trust). 2008. *Towards a Long Term Strategy for Reducing Carbon Dioxide Emissions from Our Housing Stock*. London: EST.

Faruqui, A., Sergici, S. and Sharif, A. 2010. The impact of informational feedback on energy consumption: a survey of the experimental evidence. *Energy*, **35**, pp. 1598–1608.

Gill, Z.M., Tierney, M.J., Pegg, I.M. and Allan, N. 2010. Low-energy dwellings: the contribution of behaviours to actual performance. *Building Research and Information*, **38**, pp. 491–508.

Hammond, G. and Jones, C. 2011. *Inventory of Carbon and Energy (ICE)*. Bath: University of Bath. Available at: https://www.bsria.co.uk/bookshop/books/embodied-carbon-the-inventory-of-carbon-and-energy-ice/.

Hargreaves, T., Nye, M. and Burgess, J. 2010. Making energy visible: a qualitative field study of how householders interact with feedback from smart energy monitors. *Energy Policy*, **38**, pp. 6111–6119.

Jackson, T. 2005. *Motivating Sustainable Consumption*, SDRN Briefing Note One. London: Sustainable Development Research Network. Available at: www.sd-research.org.uk/wp-content/uploads/sdrnbriefing-1motivatingsustainableconsumption_001.pdf.

Jochelson, K. 2007. The role of the government in public health: a national perspective. *Public Health*, **121**, pp. 1149–1155.

Kneifel, J. 2010. Life-cycle carbon and cost analysis of energy efficiency measures in new commercial buildings. *Energy and Buildings*, **42**, pp. 333–340.

Levine, M., Ürge-Vorsatz, D., Blok, K., Geng, L., Harvey, D., Lang, S., Levermore, G., Mongameli Mehlwana, A., Mirasgedis, S., Novikova, A., Rilling, J. and Yoshino, H. 2007. Residential and commercial buildings. In *Climate Change 2007: Mitigation: Contribution of Working Group III to the Fourth Assessment Report of the Intergovernmental Panel on Climate Change*, ed. B. Metz, O.R. Davidson, P.R. Bosch, R. Dave and L.A. Meyer. Cambridge: Cambridge University Press.

Mann, R.E., MacDonald, S., Stoduto, G., Bondy, S., Jonah, B. and Shaikh, A. 2001. The effects of introducing or lowering legal per se blood alcohol limits for driving: an international review. *Accident Analysis and Prevention*, **33**, pp. 569–583.

Martiskainen, M. 2007. *Affecting Consumer Behaviour on Energy Demand*, Final Report to EdF Energy. Brighton: Sussex Energy Group.

Meyers, R.J., Williams, E.D. and Matthews, H.S. 2010. Scoping the potential of monitoring and control technologies to reduce energy use in homes. *Energy and Buildings*, **42**, pp. 563–569.

Osmani, M. and O'Reilly, A. 2009. Feasibility of zero carbon homes in England by 2016: a house builder's perspective. *Building and Environment*, **44**, pp. 1917–1924.

Papagiannis, P., Dagoumas, A., Lettas, N. and Dokopoulos, P. 2007. Economic and environmental impacts from the implementation of an intelligent demand side management system at the European level. *Energy Policy*, **36**, pp. 163–180.

Pilkington, P. and Kinra, S. 2005. Effectiveness of speed cameras in preventing road collisions and related casualties: a systematic review. *British Medical Journal*, **330**, pp. 331–334.

RAE (Royal Academy of Engineering). 2010. *Engineering a Low Carbon Built Environment*. London: RAE. Available at: www.raeng.org.uk.

Scottish Government. 2009. *SEABS08: The Scottish Environmental Attitudes and Behaviours Survey 2008*. Edinburgh: Scottish Government. Available at: http://www.scotland.gov.uk/Publications/2009/03/05145056/11.

SDC (Sustainable Development Commission). 2007. *Lost in Transmission*. London: SDC.

Thormark, C. 2002. A low energy building in a life cycle – its embodied energy, energy need for operation and recycling potential. *Building and Environment*, **37**, pp. 429–435.

Ueno, T., Sano, F., Saeki, O. and Tsuji, K. 2005. Effectiveness of an energy-consumption information system on energy savings in residential houses based on monitored data. *Applied Energy*, **83**, pp. 166–183.

UNEP (United Nations Environment Programme). 2007. *Assessment of Policy Instruments for Reducing Greenhouse Gas Emissions from Buildings*, Report for the UNEP Sustainable Buildings and Construction Initiative. Nairobi: UNEP. Available at: http://www.unep.org/themes/consumption/pdf/SBCI_CEU_Policy_Tool_Report.pdf.

Willis, R.M., Stewart, R.A., Panuwatwanich, K., Jones, S. and Kyriakides, A. 2010. Alarming visual display monitors affecting shower end use water and energy conservation in Australian residential households. *Resources, Conservation and Recycling*, **54**, pp. 1117–1127.

Wood, G. and Newborough, M. 2003. Dynamic energy-consumption indicators for domestic appliances: environment, behaviour and design. *Energy and Buildings*, **35**, pp. 821–841.

Wood, G. and Newborough, M. 2007. Energy-use information transfer for intelligent homes: enabling energy conservation with central and local displays. *Energy and Buildings*, **39**, pp. 495–503.

Yohanis, Y.G. and Norton, B. 2002. Life-cycle operational and embodied energy for a generic single-storey office building in the UK. *Energy*, **27**, pp. 77–92.

Section C

Regulations, tools and accounting techniques

9 Regulations and incentives for LZC buildings

The practice of regulating for minimum building standards has been around for a lot longer than most people may think, and regulating for energy and emissions performance is merely the latest evolution in a long lineage of standards. Today a key question for those trying to reduce emissions from the built environment in this way is how much to use regulation and how much to use incentives. In theory at least there are many regulatory options that could achieve a significant increase in the numbers of low/zero carbon (LZC) buildings, but not all of these will be culturally, socially, economically or politically acceptable, and the acceptability of different options will vary from country to country. This poses difficult sets of questions, and the answers will vary not only internationally but also according to different subsets of the building stock, for example between domestic and non-domestic build, and by different tenure types.

This chapter provides a summary of these issues, beginning with a brief history of building standards and an overview of the current state of play, and concluding with a world-leading example of best practice.

9.1 Origins of building standards

The first recorded example of what we would now call a building code is contained within Hammurabi's Code of Laws (*ca.* 2200 BCE), which states:

229. If a builder builds a house for someone, and does not construct it properly, and the house which he built falls in and kills its owner, then that builder shall be put to death.

230. If it kills the son of the owner, the son of that builder shall be put to death.

231. If it kills a slave of the owner, then he shall pay, slave for slave, to the owner of the house.

232. If it ruins goods, he shall make compensation for all that has been ruined, and inasmuch as he did not construct properly this house which he built and it fell, he shall re-erect the house from his own means.

233. If a builder builds a house for someone, even though he has not yet completed it; if then the walls seem toppling, the builder must make the walls solid from his own means (Harper, 1904).

The Bible also contains a specific warning to house builders: 'In case you build a new house, you must also make a parapet for your roof, that you may not place bloodguilt upon your house because someone falling might fall from it' (Deuteronomy, 22:8).

It is interesting to note that these are very much, albeit basic, performance requirements and (Hammurabi's punishments aside) would not look too out of place in modern building standards. However, despite being famous for their buildings, many of which incorporated passive technologies that are being revisited in construction today, the ancient Egyptians, Greeks and Romans did not, it appears, use formal building codes, which would not reappear in history until the seventeenth century.

Although London and other British towns and cities passed various laws on building design from the end of the Dark Ages (for example, prohibiting thatched roofs in areas prone to fire) it was not until the Great Fire of London in 1666, which prompted the London Building Act of 1667, that these codes began to become legally enforceable. The 1667 Act applied to the City of London only (what we now know as the 'Square Mile') and included requirements that all buildings be built of brick or stone, and set minimum limits on street widths to help prevent the spread of fires. Yet it was not until 1774 that the Act was finally revised to cover the whole urban area, and district surveyors began keeping detailed records of their work. Elsewhere in Europe, France began regulating the construction of buildings, primarily their height, during its Second Empire (1852–70), and was followed by similar legislation in Germany and Austria.

Probably the first 'modern' building code was passed by the City of Baltimore, USA, in 1859. In 1904, the same year that the city suffered its Great Fire, the code was used as the basis for the *Handbook of Baltimore City Building Laws*, and these laws were finally appropriated as a formal building code in 1908. In 1905 the National Board of Fire Examiners (the precursor to the American Insurance Association) approved the world's first National Building Code, with Canada following suit in 1941 (Lehman and Phelps, 2005).

In the UK, Scotland was the first country to publish national building regulations, in 1963, following the Building (Scotland) Act of 1959, and they form the basis of the system still in use today.

9.2 International overview of building standards

Worldwide around 72 countries have adopted or have proposed building standards that include energy efficiency requirements. These vary according to legal status (mandatory, voluntary or mixed) and whether they pertain only to residential or non-residential buildings, or both (see Table 9.1).

Table 9.1 Overview of building standards by country

Mandatory	Mixed	Voluntary	Proposed
All buildings:			
Australia	Belgium	Egypt	Algeria
Austria	Canada	Israel	Colombia
Bulgaria	Russia	Lebanon	Cyprus
Chile	USA	Pakistan	Estonia
China		Palestine	Latvia
Czech Republic		Saudi Arabia	Malta
Denmark			Ukraine
Finland			
France			
Germany			
Hungary			
Italy			
Jamaica			
Japan			
Jordan			
Kazakhstan			
Kuwait			
Lithuania			
Luxembourg			
Netherlands			
New Zealand			
Norway			
Poland			
Portugal			
Romania			
Slovakia			
Slovenia			
South Korea			
Spain			
Sweden			
Switzerland			
Tunisia			
Turkey			
United Kingdom			
Residential only:			
Greece		Syria	
Ireland			
United Arab Emirates			
Non-residential only:			
Mexico		Cote d'Ivoire	Brazil
Singapore		Guam	Morocco
Vietnam		Hong Kong	Paraguay
		India	
		Indonesia	
		Malaysia	
		Philippines	
		South Africa	
		Sri Lanka	
		Taiwan	
		Thailand	

Source: Janda, 2009.

9.3 Mandatory and voluntary approaches for low/zero carbon buildings

A major question for those developing policies to reduce emissions from the built environment is how much to rely on voluntary approaches and how much to use regulation. Voluntary approaches, which usually rely on some form of financial incentives, are popular, because using regulation invites political and public criticism, particularly when applied to existing build. Using regulation also implicitly requires some degree of enforcement, and absorbing the additional costs that enforcement incurs. However, the problem with voluntary standards is obvious when trying to capture as many buildings as possible, and is exacerbated in periods of economic depression.

In reality these approaches exist within much wider and more complex frameworks, and may be heavily dependent on each other. Examples of regulatory approaches to new build are given in Table 9.2, examples of approaches to existing build are given in Table 9.3, and examples of voluntary approaches are given in Table 9.4.

9.3.1 Mandatory approaches

Regulation led approaches tied to mandatory standards are by far the most common, and generally effective, methods of setting energy efficiency standards for new build (Iwaro and Mwasha, 2010; Janda and Busch, 1994). Building standards tend to fall into one of two categories – prescriptive standards, which specify requirements for individual building components (walls, windows, roofs and so on), and performance standards, which address a building's performance as a whole (thermal comfort, energy efficiency, water consumption and so on). Over the last century those countries that have implemented building regulations have invariably adopted prescriptive standards, but there is now a clear move towards performance based standards, which have been found to be more effective in both improving resource efficiency and stimulating innovation and the take-up of higher performance technologies, whilst giving developers more freedom to find new ways to meet higher standards (Beerporte and Beerporte, 2007; Gann, Wang and Hawkins, 1998).

Historically the emphasis on developing building standards has been largely on new build, with the reasoning that this will drive up overall performance as existing buildings are retired. However, in many countries around the world, and particularly those with large numbers of old and/or historic buildings, low rates of turnover mean that regulating only for improving the performance of new build will be far from sufficient to meet emissions reduction targets. Therefore some more recent efforts to evolve building regulations have addressed how they can also be applied to existing buildings.

A notable example of a shift from very weak to very strong regulation is the French experience. France's rapid move towards regulating for energy efficiency began with the passing of the French BC2005 Building Standards and the RT2006 Thermal Regulation. Under current legislation all new construction must exceed the RT2006 by 15 per cent, rising to 40 per cent by 2020, and all buildings completed from the start of 2010 must achieve improvements of at least 20 per cent over the BC2005 standards. France's Agenda for Change aims to utilise technology to support the majority of all new buildings being low energy consumers or net producers of energy from 2010 onwards, supported by a combination of measures targeting new commercial and public buildings and private and public sector housing (Kitson and Daclin, 2008).

9.3.2 Voluntary approaches

Voluntary instruments for reducing emissions from buildings come in a wide range of forms, but invariably employ awareness raising and one or more of the following financial incentives:

Table 9.2 Regulatory approaches for new buildings

Policy summary	Country	Aim/targets	Supplementary policies	Strengths	Weaknesses	Evidence of success or failure
Step change in building standards supported by tax credits for improvements	France	Majority of all new build (including public and commercial sector) to be low energy or net energy producers as of 2010.	Tax credits and loans for energy efficiency and renewable energy improvements. Additional regional and national subsidies for specific measures (especially solar thermal).	Unifies prior ad hoc policies. Ambitious yet sensitive to diversity of building stock. Comprehensive package supported by existing regional policies. Strong political leadership.	Reliant on new build and further take-up of technology. Risk of loss of political will if targets are not achieved.	Current policies only recently in force so little evidence. Previous policies to increase uptake of solar thermal generally successful.
Mandatory building standards supported by stamp duty land tax relief	UK	All properties worth less than £500,000 achieving carbon neutrality to be exempt from SDLT.	Grants and loans available for energy efficiency improvements.	Simple to administer.	Reliant on consumers improving their homes.	Failure – only 24 applicants from March 2007 to January 2010.
Mandatory building standards supported by grants and loans for improvements	UK	All homes to be carbon neutral by 2016 (England), 2015 (Scotland) and 2011 (Wales).	Grants and loans available for energy efficiency improvements.	Building standards are an established policy tool and well understood by the construction industry.	Setting standards too high risks deterring investment in new build. Increasing devolution of standards increases the difficulties of compliance.	Has led to regular improvements in building performance, although turnover of stock is relatively slow.

Sources: Beerporte and Beerporte, 2007; Gann, Wang and Hawkins, 1998; Iwaro and Mwasha, 2010; Janda and Busch, 1994; Kitson and Daclin, 2008.

Table 9.3 Regulatory approaches for existing buildings

Policy summary	Country	Aim/targets	Supplementary policies	Strengths	Weaknesses	Evidence of success or failure
Generous loans (up to 100% with low interest rates) used as key driver for energy efficiency improvements. Improvements specified by an expert and must achieve minimum standards	Germany	Total refurbishment of housing stock over a 20-year period.	Mandatory building standards for new build used as minimum standards for retrofits. Some grants available but not in conjunction with loans.	Applicable to the privately rented sector, registered social landlords and housing associations. Public funding used to lever six times level of private investment. Low management costs. Government led, so seen as safe to consumers and investors.	Long term loans risk repayments being higher than savings. Loans not tied to property so must still be repaid if household moves.	Very successful – €6 billion of energy improvements installed in 2007 alone.
Interest-free loans for retrofitting	Australia	Part of national emissions reduction strategy.	Free loft insulation and grants for solar thermal.	No upfront costs. Simple to understand.	Short term (four-year) loans delay financial benefits to householders.	Recent initiative so little evidence either way.
Low interest loans for retrofitting	New Zealand	Part of national emissions reduction strategy.	Stand-alone scheme.	Simple to understand. Flexible repayment options.	Restricted eligibility. Short maximum repayment period and higher interest rates for longer repayment periods.	Recent initiative so little evidence either way.
Interest-free and low interest loans for retrofitting	France	Retrofit 400,000 homes per year from 2013, including 800,000 properties in the social sector by 2020. Reduce energy consumption of existing homes by over 38% by 2020. Train	Existing national and regional incentives, especially for solar thermal. 'Sustainable development' tax credits. Improved standards for new build.	Integrates well with other policies. Covers public and private sectors. Low interest rates and flexible repayment options.	Financial benefits depend on repayment terms. Measures not linked to achieving a minimum overall energy performance.	Recent initiative so little evidence either way.

Measure	Location	Objective	Relationship to strategy	Strengths	Weaknesses	Effectiveness
Grants for energy efficiency improvements targeted at households in, or at risk of, fuel poverty	UK	Part of UK emissions reduction strategy, but emphasis on reducing fuel poverty. 120,000 building energy professionals by 2012.	Complementary to other national and local initiatives.	Grants cover most or all of the cost of simple measures and are easy to access.	Limited scope of measures. Emphasis on alleviating fuel poverty means a portion of any investment may be used to improve comfort rather than reduce consumption.	Moderately successful – evaluation of pilot schemes found 25–45% of households lifted out of fuel poverty.
One-off council tax rebate for installing a subsidised package of energy efficiency measures	Braintree and South Cambridge councils, UK	Part of local authority emissions reduction strategies.	Complementary to other national and local initiatives.	Simple and low cost.	Very limited scope of measures (energy audit, insulation top-up, low energy light bulbs).	Limited effectiveness – Braintree's scheme had 600 applicants and insulated 250 homes for 2004–05.
Interest-free loans for installing micro-generation technologies	Kirklees, UK	Support households in installing micro-generation technologies.	As for UK nationally.	Low upfront cost to consumers. Simple to understand. Easy to adapt to different repayment models.	Limited in scope to micro-generation and energy aware households. Loans must be paid off when property is sold, so householders may not recoup investment.	A 2009 evaluation shows scheme largely achieving its aims.
Mandatory standards supported by subsidies for energy efficiency improvements	UK (proposed)	Reduce emissions from existing build by 29% by 2020. Insulate all homes by 2015. Every home to have a smart meter. 'Advanced' upgrades for 7 million homes. Minimum	Proposed package for UK, supplementary to national emissions reduction targets.	Financing options for improvements attached to property, not household. Tackles problem of inefficient rented properties. Engages and levers funding from utility companies (replacing CERT). No upfront costs. Wide range of measures covered. Includes new	Has yet to be implemented. Concerns over post-election political will in light of the recession.	N/A.

Table 9.3 Continued

Policy summary	Country	Aim/targets	Supplementary policies	Strengths	Weaknesses	Evidence of success or failure
		standards for rented properties by 2015.		plans for providing information and advice.		
Mandatory building code, levy on electricity sales, financial incentives for improvements and reducing household energy consumption	California, USA	Part of a long term strategy to reduce water and energy consumption. Ensure state level energy security.	N/A – exceeds federal level initiatives.	A 0.3¢/kWh levy on electricity is used to deliver significant energy efficiency improvements and raise further investment. Part of a long term programme so strong signals sent to construction and energy markets (e.g. new building codes being voluntary prior to becoming law). Responsive to changing conditions. Very strong and informed political leadership.	High and prolonged level of public funding. Energy crises have been key drivers, leading to very strong and cross-party political backing. May not be realistically achievable elsewhere.	Highly successful long term programme. Significant private investment used (US$230 million by 2002). Per capita energy consumption 40% below US average in 2001.
RECO – mandatory building standards apply at point of sale and major renovation, and must be met by retrofitting	Berkeley, California, USA	State level total emissions reduction target of 80% by 2050.	Wide-ranging package of measures at state level. Also Berkeley First loan scheme for solar thermal.	Strong but flexible – compliance can be achieved by the seller before sale or by the buyer within a year of sale. Applicable to owner-occupiers and landlords. Potential to adapt for use with EPCs in Europe.	Needs to be viewed within the Californian context (long history of political leadership and responding to energy crises); may be difficult to replicate outside this.	Limited evidence and difficult to disaggregate from state level legislation, but generally regarded as highly successful (in part because of capture rate). Proposed revisions will address evaluation problems.

Sources: Cooperative Bank, 2010; CSE and NEA, 2005; DECC, 2009; Dixon, Keeping and Roberts, 2008; EHA, 2009; EST, 2005, 2009; HM Government, 2010; HM Treasury, 2003; Impetus Consulting, 2009; Kirklees Council, 2009; OESD, n.d.

grants; loans; tax deductions or rebates; and labelling. Providing a complete coverage of these is outwith the scope of this book, as so many exist worldwide and many are regularly revised in line with changing policy priorities: over 30 exist in the EU alone, and Japan has had over 30,000 in operation (Lee and Yik, 2004).

Incentive schemes operate on the basis that improving the energy performance of buildings is desirable, and the main barriers are awareness and cost, so making people aware of their energy consumption and offering enough of a financial incentive for measures to reduce it should lead to them adopting it. However, the number of countries with mandatory regulations, including even Japan, is testament to the effectiveness of incentive schemes alone. The real question is how to maximise the benefits of both. Some examples of these schemes are given in Table 9.4.

9.3.3 Feed in Tariffs (FiT)

Feed in tariffs (FiTs) are a particular form of incentive scheme that pay installers of renewable energy a guaranteed price for the energy they produce, which can be electricity or heat and typically for a period of 20 years. In their most common form, as pioneered by Germany, the price is set above the market rate and commonly differentiated by the type, size and location of the technology, and prices usually decline over time as costs of generation fall. Payment is usually made by the utility company with support from government subsidies or other incentive schemes. A variation is the system developed by Spain and now being adopted elsewhere where part of the payment is determined by the hourly price for electricity, with upper and lower caps to limit the level of variation over time.

The policy community broadly agrees that 'true' FiTs must fulfil three criteria: guarantee grid access; long term energy supply contracts; and prices that provide a reasonable return above the cost of generation.

A major emerging issue with FiTs is the decisions by an increasing number of governments to make adjustments to the schemes, and for this reason any summary of their current status would be out of date within months. Adjustments are generally made to reduce the payment rates for technologies with higher than anticipated take-up rates and increase those for technologies in need of greater support. However, these adjustments can cause real confusion amongst those who have joined the schemes, and raise concerns from both potential investors and the energy industry (REN21, 2011).

9.3.4 Labelling and rating schemes

Whilst increasing requirements as part of the revision of wider building standards is the most common approach to reducing emissions, many countries are now adopting energy labelling and/or awards schemes to promote the development of low carbon buildings and building technologies. Perhaps the most well-known and widely used approach is the European Union's energy performance certificates, which became mandatory for all new buildings (and existing buildings on point of sale or change of tenancy) as of October 2008 (see Chapter 10 for more details).

Two of the most well-known rating schemes, which include an assessment of energy performance, are the Code for Sustainable Homes (England and Wales) and Energy Star for Homes™ (USA) (see Chapter 10). These approaches tend to incentivise energy and resource efficiency through ratings or 'awards' for the most efficient buildings, and vary in the scope and standard of performance being assessed; for example, the Code for Sustainable Homes Level 6 denotes a home that has been modelled to be (operationally) zero carbon.

Table 9.4 Voluntary approaches for new and existing buildings

Policy summary	Country	Aim/targets	Supplementary policies	Strengths	Weaknesses	Evidence of success or failure
Voluntary standards supported by regulation for appliances	Japan	Target is to reduce heating and air conditioning energy use by 20%.	Strict regulation of energy efficiency of appliances and consumer electronics, backed by significant investment in R&D.	Low cost. Investment strategy supports manufacturing sector. Capitalises on country strength.	Voluntary. Relatively little emphasis on buildings. Heavy reliance on technology to deliver savings.	Previous increases in standards have been largely successful.
Voluntary building energy efficiency standards supported by grants for improvements	Canada	Energy Star for Homes as a driver for installing energy efficiency measures.	Grants for improvements available through the ecoENERGY scheme.	Low financial costs for developing standards, as Energy Star is a US led programme. Acceptable to construction industry.	Voluntary. Reliant on effectiveness of incentives.	Policies are relatively ineffective – only a 5.5% decrease in average household energy consumption for 2003–07.
Voluntary standards, supported by tax incentives	USA (in general)	Energy Star for Homes as a driver for installing energy efficiency measures.	Income tax reductions for purchasing energy efficient homes. Tax rebates on improvements.	Simple mechanisms for rewarding consumers.	Voluntary. Reliant on effectiveness of incentives.	Difficult to determine owing to state-by-state adoption.

Sources: ECCJ, 2002; Geller et al., 2006.

9.4 Critical issues in regulations and incentive schemes

There are numerous barriers and opportunities to reducing emissions from both new and existing aspects of the built environment, many of which, such as user behaviour and industry culture, have wider implications and so are covered in Chapters 5 and 6. This chapter summarises three key issues that have a bearing on regulations and incentive schemes: 'hard to treat' buildings; tenure; and information and communications technology.

9.4.1 Hard to treat buildings

'Hard to treat' (HtT) buildings have been defined as existing buildings of the following kinds:

- solid wall properties;
- high rise properties;
- timber frame properties;
- properties with flat roofs;
- properties with mansard roofs;
- tenement blocks (e.g. in Scotland and parts of the USA);
- static trailers and 'holiday' or 'park' homes.

Table 6.3 specifies the technical measures to improve the carbon efficiency of HtT. However the designing of incentive schemes for these buildings faces problems that are unique to them. Labelling a building as HtT implies that there are significant technological and/or economic barriers to bringing it up to an acceptable level of performance, which in addition to the physical characteristics and condition of the building, and the economic viability of improving it, may also include social factors such as tenure.

Regulations and incentive schemes to reduce emissions from HtT properties need to be more sensitive to the problems of particular building types, or even individual buildings, as well as addressing any non-physical barriers to improvement (Roaf, Baker and Peacock, 2008).

9.4.2 Tenure

Knowing whether or not a building is owned by those who live in or use it can be an important indicator of its likely levels of energy and resource efficiency (Baker, 2007). The tenure mix of different national building stocks varies widely, and national strategies to reduce emissions invariably prioritise the most common tenures, which can leave properties occupied under other tenures lagging behind.

An example of this is the different progress on improving the performance of privately rented homes. In the UK privately rented homes constitute a small but growing proportion of the housing stock, and many of these are still suffering from the impacts of the bursting of the buy-to-let bubble in the late 1990s. This has resulted in landlords being unable or unwilling to maintain and improve their properties, and is compounded by the fact that most tenants pay their own energy bills, so the landlord can receive a return on any investment only through increased rents. However, for many landlords profit is a secondary priority to attracting good tenants who will take care of their homes and pay rent on time, rather than trying to raise their rent charges through investing in improvements (Kemp and Rhodes, 1997).

In contrast, Germany, with its high proportion of privately rented homes (49 per cent in 2001), has made significantly more progress on reducing emissions from them (Scanlon and

Whitehead, 2004). However, the German situation differs from that of many other countries in three key respects: landlords tend to invest in their properties on a very long term basis; tenancies tend to be for much longer periods; and the loans available are wide-ranging and generous. Nevertheless, the problem of 'landlord pays, tenant benefits' remains a barrier to improving energy performance that has yet to be fully overcome (Rehdanz, 2007).

Tenure can also be a barrier to improving building performance where blocks of domestic and/or commercial properties are mixed tenure, as gaining consent from all parties is usually necessary for any significant work, particularly where one or more residents are unwilling or unable to share the costs (Roaf *et al.*, 2008). Regulations and incentive schemes for low carbon buildings need to be alive to these issues.

9.4.3 Information and communications technology

The energy demand from information and communications technology (ICT) and the use of consumer electronics is having a significant impact on emissions from the built environment, and technological changes can drive emissions both upwards and downwards (Baker, 2007). An example of a new technology leading to higher emissions is the fourfold increase in energy demand for TVs and recording equipment attributed to the early stages of the UK's switch to digital TV (Karger *et al.*, 2005). This technology shift has occurred alongside a change in TV and computer monitor displays from cathode ray tubes (CRTs) to flat panel designs, but predominantly the more efficient liquid crystal displays (LCDs), which would have led to a reduction in energy demand were it not for consumers taking advantage of the dimensional benefits of the flat panel design to install TVs with larger screens (Baker and Bardsley, 2004; Russell, 2006).

At a wider scale the growth in ICT networks is having significant influences on both how much energy is being consumed by the built environment and where it is consumed. A key issue for building energy managers has been how and where best to locate server rooms in buildings that were never designed for them, whilst the growth in 'cloud computing' may pose new problems for carbon accountants in attributing emissions over organisational and international boundaries.

However, increasingly ICT is playing a more direct role in influencing energy consumption, and more sophisticated building energy management and control systems form a key part of a growing number of national emissions reduction strategies. As mentioned in Chapter 3, one technological option now gaining traction for offices and other suitable non-domestic buildings is an evolution of Edison's idea of local low voltage networks, but probably the most well-known application of ICT to reduce emissions is the smart meter, which is covered in Chapter 8.

Case study: World-leading regulation, RECO, California

In 1987 the City of Berkeley, California, passed the Residential Energy Conservation Ordinance (RECO). In terms of regulations for building energy and resource efficiency, the RECO was a landmark piece of legislation, and remains a world-leading example of best practice.

The city has set a target of reducing its greenhouse gas emissions by 80 per cent by 2050, and it intends to achieve a significant contribution to this from the residential sector. The RECO is unusual in that it addresses the problem of when and how to trigger performance improvements in existing buildings by using point of sale or transfer as the trigger for enforcing compliance with California's green building regulations. Under RECO anyone selling or transferring a property that is not already compliant with the latest Californian building regulations can either invest in the appropriate measures and demonstrate compliance (via an inspection) prior to sale or register with the city that the duty to comply will be passed on to the buyer. In the latter case the duty must be carried out within one year of the purchase and cannot be passed on to a subsequent buyer should the property change hands within this period. This allows for some flexibility in the market without the risk that some properties might slip through the net because of frequent transfers of ownership (OESD, n.d.).

Although it is difficult to disaggregate the impact of RECO from the wider state level policies, there is some evidence to suggest that the ordinance has resulted in energy efficiency improvements of around 15 per cent, as a result of which RECOs have now been implemented by other places, including San Francisco, Davis (California), Ann Arbor (Michigan), Burlington (Vermont), Boulder (Colorado) and the state of Wisconsin (Epolicy Center, n.d.). In addition a 1990 evaluation of the San Francisco RECO found that compliance was 'very high', that around 90 per cent of properties were captured as a result of being put on the market, and that the costs of implementing the ordinance were minimal – requiring the appointment of one full time officer and a part time administrator to cover the city (Vine, 1990).

References

All websites were last accessed on 30 November 2011.

Baker, K.J. 2007. Sustainable cities: determining indicators of domestic energy consumption. Ph.D. thesis, Institute of Energy and Sustainable Development (IESD), De Montfort University, Leicester.

Baker, K.J. and Bardsley, J.N. 2004. Environmental impact of LCDs and CRTs. Society for Information Display (SID) International Symposium 2004 Technical Digest of Papers. Paper presented at the SID's international symposium, Seattle, May.

Beerporte, M. and Beerporte, N. 2007. Government regulation as an impetus for innovation: evidence from energy performance regulation in the Dutch residential building sector. *Energy Policy*, **35**, pp. 4812–4825.

Cooperative Bank. 2010. Energy efficient home improvements. Available at: http://www.co-operative-bank.co.uk/bank/pdf/EnergyEfficientHomeImprovementsFactsheet.pdf.

CSE and NEA (Centre for Sustainable Energy and National Energy Action). 2005. *Warm Zones Evaluation: Final Report*. Bristol and Newcastle: CSE and NEA. Available at: http://www.warmzones.co.uk/050301%20-%20Warm%20Zones%20Evaluation%20Final%20Report.pdf.

DECC (Department of Energy and Climate Change). 2009. Boiler scrappage scheme. Available at: http://www.decc.gov.uk/en/content/cms/what_we_do/consumers/saving_energy/boiler_scheme/boiler_scheme.aspx.

Dixon, T., Keeping, M. and Roberts, C. 2008. Facing the future: energy performance certificates and commercial property. *Journal of Property Investment and Finance*, **26**, pp. 96–100.

ECCJ (Energy Conservation Centre, Japan). 2002. *Japan Energy Conservation Measures in Future*. Tokyo: ECCJ. Available at: http://www.eccj.or.jp/summary/local0303/eng/index.htm.

EHA (Existing Homes Alliance). 2009. Paying for it. Finance Working Group paper. Available at: http://www.existinghomesalliance.org/media/ExHA%20Finance%20Paper_%20Paying%20for%20it%20v4%20FINAL.pdf.

Epolicy Center. n.d. RECO factsheet. Available at: http://epolicycenter.com/media/downloads/RECO-fact_sheet.pdf.

EST (Energy Saving Trust). 2005. *Changing Climate, Changing Behaviour: Delivering Household Energy Saving through Fiscal Incentives*, July. London: EST. Available at: http://www.energysavingtrust.org.uk/uploads/documents/aboutest/fiscalupdate.pdf.

EST. 2009. *Potential for Energy Efficiency Loans for Carbon Saving Measures in Scottish Homes*. Edinburgh: EST. Available at: http://www.scotland.gov.uk/Topics/Business-Industry/Energy/Action/energy-efficiency-policy/Potentialforloans.

Gann, D.M., Wang, Y. and Hawkins, R. 1998. Do regulations encourage innovation? The case of energy efficiency in housing. *Building Research and Information*, **26**, pp. 280–296.

Geller, H., Harrington, P., Rosenfeld, A.H., Tanishima, S. and Unander, F. 2006. Policies for increasing energy efficiency: thirty years of experience in OECD countries. *Energy Policy*, **34**, pp. 556–573.

Harper, R.F. 1904. *The Code of Hammurabi King of Babylon*, 2nd edn. Chicago/London: University of Chicago Press/Callaghan and Company.

HM Government. 2010. *Warm Homes, Greener Homes: A Strategy for Household Energy Management*. London: HM Government.

HM Treasury. 2003. *Economic Instruments to Improve Household Energy Efficiency: Consultation Document on Specific Measures*. London: HM Treasury. Available at: http://www.hm-treasury.gov.uk/d/energy_efficiency.pdf.

Impetus Consulting. 2009. The case for including energy efficiency investment in the fiscal stimulus package. Report produced for Greenpeace. Available at: http://www.greenpeace.org.uk/files/EE_fiscal_stimulus_Impetus_Report.pdf.

Iwaro, J. and Mwasha, A. 2010. A review of building energy regulation and policy for energy conservation in developing countries. *Energy Policy*, **38**, pp. 7744–7755.

Janda, K.B. 2009. *Worldwide Status of Energy Standards for Buildings: A 2009 Update*. Oxford: Environmental Change Institute, Oxford University.

Janda, K.B. and Busch, J.F. 1994. Worldwide status of energy standards for buildings. *Energy*, **19**, pp. 27–44.

Karger, S., Klein, J., Reynolds, M. and Sales, B. 2005. Cost and power consumption implications of digital switchover. Report for Ofcom. Available at: http://stakeholders.ofcom.org.uk/binaries/research/tv-research/cost_power.pdf.

Kemp, P.A. and Rhodes, D. 1997. The motivations and attitudes to lettings of private landlords in Scotland. *Journal of Property Research*, **14**, pp. 117–132.

Kirklees Council. 2009. Case study: the RE-Charge Scheme. Available at: http://www.kirklees.gov.uk/community/environment/green/pdf/2009RE-ChargeCaseStudyx.pdf.

Kitson, M. and Daclin, J. 2008. Green construction standards for France. *Building Sustainable Design*, November. Available at: http://www.bsdlive.co.uk/story.asp?storycode=3126094.

Lee, W.L. and Yik, F.H.W. 2004. Regulatory and voluntary approaches for enhancing building energy efficiency. *Progress in Energy and Combustion Science*, **30**, pp. 477–499.

Lehman, J. and Phelps, S. (eds). 2005. *West's Encyclopaedia of American Law*. Eagan, MN: West.

OESD (Office of Energy and Sustainable Development), City of Berkeley, n.d. *Residential Energy Conservation Ordinance*. Berkeley, CA: OESD. Available at: http://www.ci.berkeley.ca.us/ContentDisplay.aspx?id=16030.

Rehdanz, K. 2007. Determinants of residential space heating expenditures in Germany. *Energy Economics*, **29**, pp. 167–182.

REN21 (Renewable Energy Policy Network for the 21st Century). 2011. *Renewables 2011: Global Status Report*. Paris: REN21. Available at: http://www.ren21.net/Portals/97/documents/GSR/GSR2011_Master18.pdf.

Roaf, S., Baker, K. and Peacock, A. 2008. *Evidence on Hard to Treat Properties*. Edinburgh: Scottish Government. Available at: www.scotland.gov.uk/Publications/2008/10/17095821/0.

Russell, B. 2006. Flat screen televisions 'will add to global warming'. *Independent*, 1 November. Available at: http://environment.independent.co.uk/climate_change/article1945758.ece.

Scanlon, K. and Whitehead, C.M.E. 2004. Housing tenure and mortgage systems: how the UK compares. *Housing Finance*, **64**, pp. 41–51.

Vine, E. 1990. *Building Code Compliance and Enforcement: The Experience of San Francisco's Residential Energy Conservation Ordinance and California's Building Standards for New Construction*. Berkeley, CA: Lawrence Berkeley National Laboratory. Available at: http://www.escholarship.org/uc/item/0b46z9tq?display=all.

10 Tools and assessment systems for the built environment

The previous chapter presented an overview of regulation and incentives for LZC buildings. While the focus of Chapter 9 was on those schemes creating an enabling environment for low/zero carbon energy for use in buildings, this chapter turns its attention to built environment focused tools and systems that aim to improve the carbon efficiency of building fabric as well as the urban context in which buildings are situated. The chapter first focuses on assessment systems for carbon in buildings as a 'product' (i.e. the building fabric and its operational energy requirements), before presenting tools and assessment systems for buildings embedded in a 'system' (i.e. urban carbon assessment incorporating the assessment of emissions from all the activities across the urban area, including buildings' energy use, industrial processes and transportation). Readers are referred to Chapter 8 for details on the assessment of carbon embodied in buildings, from the extraction of raw materials and manufacturing through to its use and final reuse, recycling or disposal.

10.1 Tools for building carbon assessment

Building carbon emission tools and assessment systems typically focuses on operational energy requirements and associated carbon emissions. The focus on carbon is an offshoot of the nearly 40-year history of building energy assessment, the latter part of which has seen an explosion of new tools, modelling techniques and assessment systems. The wider sustainability assessment of buildings, though relatively new (less than 20 years old), has seen an even faster growth of tools. In the field of urban sustainability assessment alone, Walton *et al.* (2005) found over 675 tools! Most building sustainability assessment tools measure energy use (and therefore by extension carbon emission (as can be seen in Table 10.1), but systems that specifically assess carbon in buildings remain a relatively new phenomenon.

10.1.1 ISO standards

The International Organization for Standardization (ISO) has produced various standards for buildings and, whilst a full coverage of these is beyond the scope of this chapter, two are of particular relevance:

*   ISO 21930:2007 (Sustainability in Building Construction – Environmental Declaration of Building Products) addresses the sustainability of individual building components, allowing developers to identify those components that have the lowest environmental impacts across their entire lifecycles (ISO, 2007).
*   ISO 21931-1:2010 (Sustainability in Building Construction – Framework for Methods of Assessment of the Environmental Performance of Construction Works – Part 1: Buildings) is aimed at improving the environmental performance of buildings by providing an internationally agreed framework for methods used in assessing their environmental stewardship.

Table 10.1 Issue coverage by key international building sustainability assessment tools

Tool name	Country of origin	Start year	Sustainable sites	Land use and ecology	Energy efficiency	Water efficiency	Materials and resources	Environmental loading	Indoor environment	Service quality	Transport	Social aspects	Economic aspects	Cultural aspects	Regional priority	Management	Innovative design	Awareness and education	Lifecycle assessment
BREEAM	UK	1990		X	X	X	X	X	X		X	X					X		
GB/SB Tool	International	1996	X		X	X	X	X	X	X		X		X					
LEED	USA	1998	X		X	X	X	X	X		X				X		X	X	X
Green Globe	Canada/ USA	2000	X	X	X	X	X	X	X						X				
CASBEE	Japan	2001	X	X	X		X	X	X	X									
Green Star	Australia	2003		X	X	X	X	X	X	X	X	X					X		
LEED – Canada	Canada	2004	X				X	X	X	X							X		
HQE	France	2004	X	X	X	X	X	X	X	X		X				X			
LiderA	Portugal	2005	X	X	X	X	X	X	X	X		X					X		X
Green Mark	Singapore	2005	X		X	X			X		X						X		
Verde	Spain	2005	X	X	X	X	X	X	X	X		X	X			X			
DGNB	Germany	2009	X	X	X	X	X	X	X	X	X	X	X	X	X	X	X	X	X
GBI	Malaysia	2009	X		X	X	X		X								X		
Pearl	Abu Dhabi	2010	X	X	X	X	X	X	X	X	X	X	X	X	X	X	X	X	X

Source: Poston, Emmanuel and Thomson, 2010.

Note: See Poston *et al.*, 2010 for full references to tool websites.

The standard is an attempt to bridge the gap between national and international standards by providing a common framework for energy and sustainability assessment throughout the lifecycle of a building. ISO 21931 is intended to be used in conjunction with, and following the principles set out in, the ISO 14020 group of international standards on environmental labelling, as well as ISO 14040 on lifecycle assessment and ISO 15392 on general principles of sustainability in building construction (ISO, 2010).

The ISO is also working to achieve greater harmony and comparability between different building codes or regulations through its Building Code Effectiveness Grading Schedule (BCEGS®). The Schedule rates building codes on a scale of 1 to 10, with 1 representing exemplary commitments to enforcement and 10 representing no recognisable enforcement. Insurers can use these gradings to offer premium rates for those buildings constructed under the most strictly enforced codes (ISO, 2011).

10.1.2 EU Energy Performance Certificates (EPCs)

In the EU the production of energy performance certificates (EPCs) for all new buildings and those undergoing major renovation became mandatory in 2007, and in 2008 the requirement was extended to rented properties upon any change of tenancy. The labels were originally introduced under the EU's 2002 Directive on the Energy Performance of Buildings (2002/91/EC), and must be displayed in all buildings covered by the scheme, and in the case of public buildings must be located so as to be visible to all building users.

The labels rate buildings for energy efficiency and environmental impact (CO_2 emissions) on a scale of A to G, and also show the potential rating the building could achieve if renovated to the latest standards. These labels have the benefits of being consistent across Europe and appearing very similar to the EU's energy efficiency labels for appliances, which should enhance public awareness and understanding. In 2010 the Directive (now Directive 2010/31/EU) was recast to strengthen the energy performance requirements and streamline and clarify some of its provisions (EC, 2011).

10.1.3 Energy Star

Energy Star is a set of energy assessment and labelling tools developed by the US Environmental Protection Agency (EPA), appropriate for new dwellings and many (but not all) new and existing non-domestic buildings.

Energy Star for Homes is an assessment that awards the Energy Star label to those achieving an energy efficiency performance at least 15 per cent higher than those built to the standards set out in the 2004 International Residential Code (IRC). To gain the label a developer must partner with the EPA and submit design proposals for the dwelling(s), and then work with an approved EPA assessor to implement these through to completion and final assessment. An element of flexibility is allowed for by assessors having the option to use a standard list of measures, or to opt for a more customised approach that must be verified with appropriate modelling software. Unlike many other energy efficiency assessments, Energy Star for Homes includes the requirement for developers to install energy efficient lighting and appliances, which must themselves achieve an Energy Star label (EPA, 2011a).

Energy Star for Buildings is the non-domestic equivalent of the homes scheme. Aside from manufacturing plants, which require the use of a separate energy performance indicator tool, those managing buildings eligible for the scheme can generate a rating on a scale of 1 to 100 using

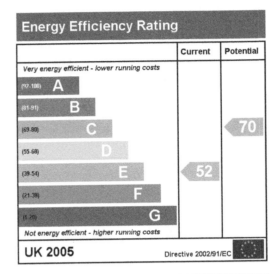

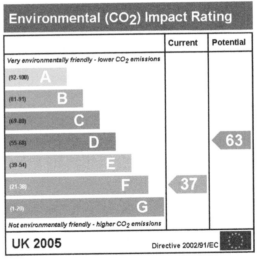

Figure 10.1 An example of an energy performance certificate

the EPA's free online Portfolio Manager tool. Those that achieve a score of 75 or higher, which must be verified by an independent assessor, can then be awarded the Energy Star label.

One of the potential strengths of the Energy Star assessment is that it allows for energy performance to be assessed with respect to the age and potential performance of the building, so a new building constructed to meet minimal energy efficiency standards should receive a lower rating than a comparable older building that has been renovated to improve energy performance, even if the latter still consumes more energy per unit area. Those buildings achieving the label should consume around 35 per cent less energy than average. This emphasis on performance improvement was reflected in the award of the first label to a 17-year-old, 74,000-square-foot municipal office building in San Diego, California, in 1999. Another strength of the scheme is that it capitalises on the familiarity and awareness of the Energy Star label, which used to be a common sight on the loading screens of many desktop computers (EPA, 2010, 2011b).

10.1.4 UK National Home Energy Rating (NHER) Scheme

The UK's National Home Energy Rating Scheme was initiated in 1990 and is owned and operated by National Energy Services (NES). The scheme offers a set of domestic energy assessment tools, which are based on the outputs from the Milton Keynes Energy World Project, and increase in complexity from Level 0 upwards. Dwellings are rated on a scale of 0 to 20, with 20 equating to net zero CO_2 emissions and net zero running costs. For comparison, a new dwelling built to the UK's 2006 building regulations would achieve a score of 10 (NES, 2011).

NHER is a precursor to the more widely used (and less complex) Standard Assessment Procedure (SAP), and is arguably both more accurate and more flexible for three key reasons. First, NHER allows for the actual location of the dwelling, whereas SAP assumes the dwelling to be located at a latitude equivalent to the East Pennines (northern England). Secondly, NHER allows for the input of an actual heating regime, whereas SAP uses a standard assumption. Finally, and most critically, NHER also allows for the input of the number of occupants, whereas SAP again uses an assumption, making NHER more sensitive to human influences on domestic energy consumption.

This added complexity also enables greater flexibility for researchers untangling underlying and emerging influences on domestic energy consumption to adapt the assessment for these needs, for example to demonstrate the significance (by proxy) of changes in household behaviour on annual energy consumption (Baker, 2007).

Despite the relative dominance of SAP in the UK, in Scotland NHER has recently enjoyed a renaissance period, with both SAP and NHER ratings used in policy development. However, at the time of writing the scheme faces an uncertain future (Baker and Emmanuel, 2011).

10.1.5 Code for Sustainable Homes

The UK's Code for Sustainable Homes is a voluntary standard introduced in 2007 and became mandatory for all new homes in England and Wales in 2008. The Code, which is based on and replaces the Ecohomes rating system developed by the UK's Buildings Research Establishment (BRE), rates dwellings on a scale of 1 to 6, with 6 including achieving net zero operational emissions. To gain a rating at each level the dwelling must satisfy requirements under eight criteria: energy and CO_2 emissions, water and surface water run-off, materials, waste, pollution, health and well-being, management, and ecology. Although originally designed for single dwellings the Code has been adapted to allow for the assessment of multiple-occupancy buildings such as halls of residence.

As of 2010 all new private and public housing must achieve a minimum of Level 3, in accordance with the requirements of the 2010 building regulations for England and Wales (DCLG, 2010). Scotland, which gained the power to set its own building regulations following devolution, has developed a separate labelling scheme, and Wales may follow suit, having gained the same power in 2011.

10.1.6 Scottish Sustainability Label for Domestic Buildings

The Scottish Sustainability Label for Domestic Buildings was added to the Scottish building standards in 2011. The label aims to reward developers who go beyond the energy efficiency requirements set out in the Scottish building regulations, and includes assessment of wider sustainability improvements such as natural lighting, resource efficiency, noise insulation, enhanced disabled access, and provision for home offices and sustainable transport.

The label rates dwellings on a Bronze to Platinum rating across eight categories – CO_2 emissions, energy for space heating, energy for water heating, water efficiency, well-being and security, material use and waste, flexibility and adaptability, and optimising performance – with the dwelling needing to meet the criteria for all eight categories in order to attain the award at each level. Those that comply with the existing building regulations automatically achieve the Bronze award. An additional 'Active' award can be gained at each level by the use of low and zero carbon generating technologies (LZCGTs) such as micro renewables.

Although the emissions category does require modelling, the other seven categories are assessed using a reporting format similar to those used for planning applications, thereby limiting the additional work required on the part of developers. A non-domestic version of the label is under development (Scottish Government, 2011).

10.2 Tools and techniques for urban carbon management

A key to understanding and quantifying urban carbon emissions is to treat urban processes as 'metabolism' (Wolman, 1965). Urban metabolism can be interpreted either primarily in terms of

Figure 10.2 Scottish Sustainability Label for Domestic Buildings

Source: Scottish Government, 2011.

energy flows or more broadly including a city's flows of water, materials and nutrients (Kennedy *et al.*, 2009). However, a key issue is the lack of reliable, published data on comprehensive energy use in cities. It is often the case that most cities have data for urban transportation and electricity use.

Accounting for urban carbon emission is very tricky. Given the large hinterland needed to support most cities, GHG emissions by or for cities do not always occur within their political boundaries, and therefore data sourcing is extremely complicated. Cities act as points of convergence, and although they do not always control the totality of emissions made on their behalf they do have a say in the emissions by way of consumption, legislation and even culture (Lebel *et al.*, 2007).

Lebel *et al.* (2007) suggested four broad categories of emissions associated with cities:

- locally produced and locally consumed – 'direct' emissions;
- locally produced but consumed elsewhere – 'responsible' emissions;
- produced elsewhere but locally consumed – 'deemed' emissions;
- produced and consumed elsewhere – 'logistic' emissions (see Figure 10.3).

Of these, cities have great influence on both direct and deemed emissions through policy options as well as cultural influence (Lebel *et al.*, 2007).

10.2.1 Developing a city carbon budget

A key question to be answered in developing a city carbon budget is assigning local 'blame' for emissions: how much of the total emissions reduction responsibility should be placed on localities (Salon *et al.*, 2010) and what form the penalties for non-compliance should take. This will depend on the governance structure of the city and its relationship to the surrounding region and the nation. Typically nations will set the rules, budget allocations and finances for carbon reduction and provide methodologies for the emission allocations (with the help of the tools described

Consumption		
	Local	*Elsewhere*
Local	**Direct emissions** Local emissions from *consumption* and *production* assigned to the city	**Responsible emissions** Local emissions from *production* assigned to the city
Elsewhere	**Deemed emissions** Local emissions from *consumption* assigned to the city	**Logistic emissions** Local emissions from transit between *consumption* and *production* via a city is assigned to the city

(Production)

Figure 10.3 Accounting for urban carbon

Source: Based on Lebel *et al.*, 2007.

below or protocols developed elsewhere). Cities will then be responsible for developing local initiatives to reduce their emissions within the carbon budget and implement these initiatives.

Budget allocation methods

There are potentially four budget allocation methods:

- allowance allocation via auction;
- uniform allowance allocation on a per capita basis;
- using current per capita emissions as a starting point and moving gradually to a uniform allowance allocation on a per capita basis;
- using current per capita emissions as a starting point and reducing the allowance allocation by the same percentage for all localities (Salon *et al.*, 2010).

It is important to keep in mind the equity dimensions of a budget, both a procedural equity (who determines the budget and who implements it) and a consequential equity (whose actions lead to whose emissions and their reductions). It is necessary to ensure both that the city carbon budget is transparent and that it does not stifle economic growth.

Determining sectoral emissions

Key to the success of a city carbon budget is the baseline setting in terms of sectoral emissions and assigning these correctly to a city. As shown in Chapter 7, the main sources of urban emission are buildings, electricity supply, transport and waste. Of these the waste sector represents the least significant source and may be avoided; however, local realities of a particular city might warrant its inclusion.

BUILDINGS

Since most of the emissions associated with building energy use come from electricity and gas or home heating fuel use, it will be key to quantify electricity emissions using appropriate emission factors. Many of the electricity, gas and heating fuel utilities are consolidated industries that may make it easy to obtain the relevant data from a handful of companies. It may be appropriate to use national or regional emission factors (discussed in Chapters 11 and 12) to these data to quantify the emissions, since power generation is often a centralised activity.

Because newly constructed buildings are generally more energy efficient than older buildings, there is a potential equity differential between localities experiencing fast growth and those that are stable or declining. If total building emissions per capita is the metric used to determine compliance with city carbon budgets, fast growing cities might be able to meet their buildings sector budget without taking local action. This would happen if there is enough new construction (with associated mandated efficiency levels) so that, on a per capita basis, average emissions would come down even without local action.

TRANSPORT

On-road vehicles move freely between localities, emitting greenhouse gases as they go. The best method of assigning these emissions to localities and measuring them is not immediately obvious but should be based upon some measurement of distance travelled (vehicle kilometres travelled – VKT; see Table 10.2) by vehicles in that region. The ideal assignment methodology should:

Table 10.2 VKT assignment and measurement methods

VKT assignment	VKT measurement
1 VKT within locality	Loop detector data, model
2 VKT by refuelling in locality	Fuel sales, average fuel economy
3 VKT by vehicle home locality	Odometer readings
4 ½ VKT by vehicle origins in locality	
½ VKT by vehicle destinations in locality	Travel survey, model
5 VKT by vehicle home locality adjusted for new non-residential development	Odometer readings, survey of visitors to new non-residential development

Source: Salon *et al.*, 2010.

- enable precise local travel measurement;
- maximise options for local government action to reduce the assigned distance travelled;
- avoid encouraging local policy that might actually increase distance travelled at a regional level.

OTHER EMISSIONS

Local governments control policy levers that affect greenhouse gas emissions outside of these base emissions categories. Therefore a mechanism could be included in a city carbon budgets policy, that allows localities to adjust their base emissions if they have reduced emissions in another area. Examples of such actions include reducing emissions from local government operations, promoting transportation technologies above and beyond the state or national requirements, or promoting lower carbon footprint (embodied emissions) building materials. For these 'extra-base' activities, the burden would be on the locality to measure the actual reduction in emissions, using an approved measurement methodology.

Banking, borrowing and trading emissions

Many low carbon initiatives will yield emission reductions only in the long term; therefore it may be necessary to allow for local flexibility in emissions. One provision that would create this temporal flexibility is the banking and borrowing of emissions allowances. An emissions allowance is an authorisation to emit a certain amount of a pollutant, in this case greenhouse gases. Within the city carbon budgets framework, each locality would be given emissions allowances equal to its budget for each year. With allowance banking, a locality could save part of its allocated emissions budget for use at a later time. Specifically, if a locality emits fewer greenhouse gases than it is allowed in one period, it can 'bank' the difference, allowing higher emissions in future periods than would otherwise be allowed. Allowance borrowing is the reverse concept – if a locality's emissions are greater than its budget in one period, it could 'borrow' allowances from a future period's budget to make up the difference. Allowance banking could be unlimited. However, there should be limits on allowance borrowing, since budgets are likely to be designed to fall over time, and a large allowance 'debt' would become difficult to pay back (Salon *et al.*, 2010).

Penalties and rewards

The purpose of penalties and rewards ('carrots' and 'sticks') is to make city carbon budgets compulsory as well as financially rewarding. One without the other will only make the scheme less effective.

One reward possibility suggested by Salon *et al.* (2010) is that of a carbon trust fund created from a portion of the funds that may come from the auctioning of greenhouse gas emissions allowances under an industry cap and trade programme. Another financial mechanism that could be used to encourage compliance is allocation of state and national transportation funds. If localities fail to meet their target budgets in the first years of the programme, but are clearly experimenting with local initiatives that aim to reduce greenhouse gas emissions, then penalties are not in order. As experience accumulates with city carbon budgets, we will gain a better understanding of which types of initiatives are likely to be successful in which types of communities, and how much they cost to implement. Along with this knowledge comes greater local responsibility. If localities continue to miss their targets under this funded mandate after it is clear what they need to do to achieve them, then penalties should begin to apply. These could take the form of either withheld transportation funds or direct fines (Salon *et al.*, 2010).

Timing

Implementation of a city carbon budgets programme could occur in three stages. The first stage would be voluntary adoption by localities of non-binding carbon budgets. Local governments could receive technical assistance from higher levels of government, but would not be eligible for financial implementation assistance, because these budgets would be non-binding. The second stage would be voluntary adoption of a legally binding budget. Local governments could receive both technical and financial assistance, both to support compliance with the budgets and to encourage adoption of budgets. The third stage would be the full policy framework: mandatory adoption of budgets by all local governments, with accompanying technical and financial assistance from the state or nation (Salon *et al.*, 2010).

10.2.2 Long-range Energy Alternatives Planning (LEAP) system

While strictly not an 'urban' tool, the LEAP system is a decision support software developed by the Stockholm Environment Institute (Heaps, 2008) for energy policy analysis and climate change mitigation assessment for cities, regions and even countries. It is used to track energy consumption, production and extraction in all sectors of an economy, and the protocols include both energy and non-energy sector GHGs.

LEAP's modelling capabilities operate at two basic conceptual levels: in-built calculations for all of the 'non-controversial' energy, emissions and cost–benefit accounting calculations and user specified time-varying data or multi-variable models (Heaps, 2008). It is designed around long range (typically 20–50 years) scenario analysis. These scenarios specify how an energy system (such as energy supply, transport, building and so on) might evolve over time. Using LEAP, policy analysts can create and then evaluate alternative scenarios by comparing their energy requirements, their social costs and benefits and their environmental impacts.

Cities are beginning to make use of the LEAP model to plan their carbon reduction strategies. Recent examples include Copenhagen (City of Copenhagen, 2009), Cape Town (Winkler *et al.*, 2005) and Seattle (City of Seattle OSE, 2011). The key to using LEAP at a city level (as opposed to a national or regional level) is to set the boundary conditions and responsibilities correctly. More details on boundaries and other accounting principles are given in Chapter 11. A more focused approach to using LEAP in cities should ensure all the conditions (including responsibility for emission, vehicle kilometres travelled and so on) specified in 10.2.1 above are met. This will especially be the case with respect to electricity generation and fuel use (only those that are 'consumed' within the city borders should be included) and transport energy (using the VKT protocols specified in Table 10.2). Care should also be exercised in accounting for the electricity used in district heating systems in cities where such systems are prevalent.

10.2.3 International Local Government GHG Emissions Analysis Protocols (IEAP)

The IEAP is a carbon accounting system designed specifically for cities to inventory their emissions, set targets to reduce them and monitor and report progress towards achieving the targets. IEAP is based on:

- *relevance:* GHG inventory to reflect local emissions over which local control exists;
- *completeness:* attempt to include all GHGs within a chosen boundary;
- *consistency:* facilitate meaningful comparisons of emissions over time;
- *transparency:* all relevant issues addressed and all assumptions disclosed;
- *accuracy:* sufficient to enable users to make decisions with reasonable quality assurance (ICLEI, 2009).

IEAP accounts for six GHGs (CO_2, CH_4, N_2O, PFCs, HFCs and SF_6) according to their global warming potential (see Table 10.3). Two kinds of boundaries are recognised: organisational boundary (functions directly under local government control) and geopolitical boundary (physical area or region over which a local government has jurisdictional authority). Emissions of the former are referred to as 'government operation emissions', and emissions of the latter are referred to as 'community emissions'. Table 10.3 shows how these two boundaries map on to the UNFCCC's macro-sectors.

The reporting requirements of IEAP are specified according to the boundary (government or community) and the scope of emissions (see Chapter 11 for further details on boundary conditions and emission scopes).

Table 10.3 IEAP sectors relative to macro-economy sectors according to the UNFCCC

UNFCC's macro-sector		IEAP government emissions	IEAP community emissions
Energy	Stationary	Buildings and facilities	Residential
		Street lighting and traffic signals	Commercial
		Water and wastewater treatment, collection and distribution (energy only)	Industrial
	Transport	Government transport Employee commute	Transportation
	Fugitive emissions	Other	Other
Industrial processes		Other	Other
Agriculture		Other	Agricultural emissions/other
Land use, land use change and forestry		Other	Other
Waste	Solid waste disposal Biological treatment Incineration and open burning Wastewater treatment and discharge	Waste	Waste

Source: Derived from ICLEI, 2009.

10.2.4 Local authority based tools

What follows is a brief presentation of some local authority based approaches to 'community' inventories at the city scale. Community emissions are those that occur within city boundaries (see 10.2.3 above and Table 10.3 for a distinction between 'community' and 'government' emissions).

Cities for Climate Protection (CCP) software

CCP is designed to help cities that are part of the International Council for Local Environmental Initiatives (ICLEI)'s Cities for Climate Protection Campaign to develop their local climate action plans. CCP develops GHG emission inventories from energy use and waste generation and can quantify financial savings, air pollutant reductions and other co-benefits (available at: www.torriesmith.com).

REAP

REAP provides baseline data and scenario modelling of carbon, greenhouse gas and ecological footprints for the regions and local authorities of the UK. The scenario functions in REAP enable a policy maker to answer 'What if' questions about the effects of policy on the environment, helping to formulate strategies for local, regional and national government (software available at: www.resource-accounting.org.uk/reap).

DREAM-City

DREAM-City is based on the Dynamic Regional Energy Analysis Model (DREAM) to calculate energy and emissions from the urban domestic, services, industrial and transport sectors. It is based on the STELLA program and produces energy demand in each of the above sectors in terms of different fuels. It has been validated in the UK (Titheridge, Boyle and Fleming, 1996) and is freely available.

Energy and Environmental Prediction (EEP) model

The EEP model is a decision support tool more widely used for sustainable development in cities. It is based on a geographic information system (GIS) platform and is capable of accounting for production and consumption related emissions in the domestic and non-domestic buildings, transport and industrial sectors. More details are given in Jones, Williams and Lannon (2000).

References

All websites were last accessed on 30 November 2011.

Baker, K.J. 2007. Sustainable cities: determining indicators of domestic energy consumption. Ph.D. thesis, Institute of Energy and Sustainable Development (IESD), De Montfort University, Leicester.
Baker, K.J. and Emmanuel, R. 2011. Personal communications with the Building Standards Division, Scottish Government, as part of research activities. Some documents confidential to client.
City of Copenhagen. 2009. *Copenhagen Climate Plan.* Copenhagen: City of Copenhagen. Available at: www.kk.dk/climate.
City of Seattle OSE (Office of Sustainability and Environment). 2011. *Getting to Zero: A Pathway to a Carbon*

Neutral Seattle. Seattle, WA: City of Seattle OSE. Available at: http://www.seattle.gov/environment/documents/CN_Seattle_Report_May_2011.pdf.

DCLG (Department for Communities and Local Government). 2010. *Code for Sustainable Homes: Technical Guide – 2010.* London: Department for Communities and Local Government. Available at: http://www.communities.gov.uk/publications/planningandbuilding/codeguide.

EC (European Commission). 2011. *Energy Efficiency in Buildings.* Brussels: EC. Available at: http://ec.europa.eu/energy/efficiency/buildings/buildings_en.htm.

EPA (US Environmental Protection Agency). 2010. *A Decade of Energy Star Buildings: 1999–2009.* Washington, DC: EPA. Available at: http://www.energystar.gov/ia/business/downloads/Decade_of_Energy_Star.pdf.

EPA. 2011a. *How New Homes Earn the Energy Star.* Washington, DC: EPA. Available at: http://www.energystar.gov/index.cfm?c=new_homes.nh_verification_process.

EPA. 2011b. *The Energy Star for Buildings and Manufacturing Plants.* Washington, DC: EPA. Available at: http://www.energystar.gov/index.cfm?c=business.bus_bldgs.

Heaps, C. 2008. *Long range Energy Alternatives Planning System: An Introduction to LEAP.* Stockholm: Stockholm Environment Institute. Available at: http://www.energycommunity.org/documents/LEAPIntro.pdf.

ICLEI (International Council for Local Environmental Initiatives). 2009. *International Local Government GHG Emissions Analysis Protocol (IEAP),* version 1.0. Bonn: ICLEI. Available at: www.iclei.org/fileadmin/user_upload/documents/Global/Progams/CCP/Standards/IEAP_October2010_color.pdf.

ISO (International Organization for Standardization). 2007. *New ISO Standard Encourages Sustainability in Building Construction.* Geneva: ISO. Available at: http://www.iso.org/iso/pressrelease.htm?refid=Ref1344.

ISO. 2010. *ISO Standard Set to Reduce Environmental Impact of Buildings.* Geneva: ISO. Available at: http://www.iso.org/iso/pressrelease.htm?refid=Ref1344.

ISO. 2011. *Building Code Effectiveness Classifications.* Geneva: ISO. Available at: http://www.iso.com/Products/Building-Code-Effectiveness-Classifications/Building-Code-Effectiveness-Classifications.html.

Jones, P., Williams, J. and Lannon, S. 2000. Planning for a sustainable city: an energy and environment prediction model. *Journal of Environmental Planning and Management,* **43**, pp. 855–872.

Kennedy, C., Steinberger, J., Gasson, B., Hansen, Y., Hillman, T., Havránek, M., Pataki, D., Phdungsilp, A., Ramaswami, A. and Mendez, G.V. 2009. Greenhouse gas emissions from global cities. *Environmental Science and Technology,* **43**, pp. 7297–7302.

Lebel, L.A., Garden, P., Banaticla, M.R.N., Lasco, R.D., Contreras, A., Mitra, A.P., Sharma, C., Nguyen, H.T., Ooi, G.-L. and Sari, A. 2007. Integrating carbon management into the development strategies of urbanizing regions in Asia. *Journal of Industrial Ecology,* **11**, pp. 61–81.

NES (National Energy Services). 2011. *What Is an NHER Rating?* Milton Keynes: NES. Available at: http://www.nesltd.co.uk/blog/what-nher-rating.

Poston, A., Emmanuel, R. and Thomson, C. 2010. Developing holistic frameworks for the next generation of sustainability assessment methods for the built environment. In *Proceedings of the 26th Annual ARCOM Conference, Leeds,* vol. 2, ed. C. Egbu, pp. 1487–1496. Reading: Association of Researchers in Construction Management (ARCOM).

Salon, D., Sperling, D., Meier, A., Murphy, S., Gorham, R. and Barrett, J. 2010. City carbon budgets: a proposal to align incentives for climate-friendly communities. *Energy Policy,* **38**, pp. 2032–2041.

Scottish Government. 2011. *Technical Handbooks 2011: Domestic Sustainability.* Edinburgh: Scottish Government. Available at: http://www.scotland.gov.uk/Topics/Built-Environment/Building/Building-standards/publications/pubtech/thb2011domsust.

Titheridge, H., Boyle, G. and Fleming, P. 1996. Development and validation of a computer model for assessing energy demand and supply patterns in the urban environment. *Energy and Environment,* **7**, pp. 29–40.

Walton, J.S., El-Haram, M., Castillo, N.H., Horner, R.M.W., Price, A.D.F. and Hardcastle, C. 2005. Integrated assessment of urban sustainability. *Engineering Sustainability,* **158**, pp. 57–66.

Winkler, H., Borchers, M., Hughes, A., Visage, E. and Heinrich, G. 2005. *Cape Town Energy Futures: Policies and Scenarios for Sustainable City Energy Development.* Cape Town: University of Cape Town, Energy Research Centre. Available at: http://www.energycommunity.org/documents/CapeTownEnergy.pdf.

Wolman, A. 1965. The metabolism of cities. *Scientific American,* **213**, pp. 179–190.

11 Carbon, GHG and sustainability accounting

'Not everything that can be counted counts, and not everything that counts can be counted' (attributed to Einstein). *But* what can be counted can be *managed* – so it's important to get things right.

Carbon accounting is one of the most complex and rapidly evolving fields in sustainable development today. Increasingly, the outputs of carbon accounting are being used to inform decision making at all levels of society – from evaluating low carbon community projects through to policy and investment decisions for multinational companies and international bodies such as the United Nations and the World Bank.

Whilst the scopes and complexities of carbon assessments (or 'carbon footprints') will vary widely, the fundamental principles and methods should share the common framework that has emerged over recent years. This variety and the speed at which progress is being made in the field mean that it would not be useful to describe a specific methodology in depth, as none is able to offer a one-size-fits-all solution to any given set of requirements. However, there are key aspects of carbon accounting that should be understood and used by everyone working in the field or needing to understand its outputs.

Carbon accounting is just one tool in our arsenal in the fight against climate change, and a fairly blunt one too, but it is important to remember that, whilst decisions made as the result of carbon accounting exercises may not always be the best ones, the real value is in ensuring that we don't make the wrong ones.

11.1 International standards – the GHG Protocols

The most widely used and respected framework for carbon accounting is that set by the GHG Protocols group, which has been developed as a partnership between the World Resources Institute (WRI) and the World Business Council for Sustainable Development (WBCSD) and is based on over ten years of work in the field. The work underpins a huge range of initiatives on carbon accounting, including the ISO standards, the Climate Registry, and the carbon accounting practices of many governments and businesses around the world. Thus the documents and tools provided by the group should be a primary reference for anyone involved in carbon accounting.

Most importantly, the GHG Protocols set out the five principles of carbon accounting, which should be used as part of the development and review of all new and existing methodologies and tools.

11.1.1 The five principles of carbon accounting

- *Relevance:* Ensure the GHG inventory appropriately reflects the GHG emissions of the organisation or product, and serves the decision-making needs of the users – both internal and external to the organisation.
- *Completeness:* Account for and report on all GHG emissions sources and activities within the chosen inventory boundary. Disclose and justify any specific exclusions.
- *Consistency:* Use consistent methodologies to allow for meaningful comparisons of emissions over time. Transparently document any changes to the data, inventory boundary, methods, or any other relevant factors in the time series.
- *Transparency:* Address all relevant issues in a factual and coherent manner, based on a clear audit trail. Disclose any relevant assumptions and make appropriate references to the accounting and calculation methodologies and data sources used.
- *Accuracy:* Ensure that the quantification of GHG emissions is systematically neither over nor under actual emissions, as far as can be judged, and that uncertainties are reduced as far as practicable (WRI, 2004).

In theory at least, any two carbon footprints can be compared according to these criteria, even if aspects such as the boundaries, assumptions and conversion factors are different. Therefore if two carbon footprints for an organisation or product (etc.) produce different results, understanding how and why their methodologies differ with respect to these principles should provide sufficient evidence as to which is the more relevant for the intended purpose.

In addition, it is necessary to understand how complete a carbon footprint is in order to identify any omissions and/or cases of double counting. Depending on the purpose and complexity of an individual carbon footprint it may be unavoidable or unnecessary to produce a totally 'complete' footprint. For example, an organisation seeking to justify significant investment in installing a renewable energy technology in its headquarters may not to wish to bear the cost of a full organisational assessment. However, if that technology is to be used to charge electric vehicles there is an obvious benefit in developing a complete picture of how those vehicles are used and how much electricity will be contributed from renewables. Another example, likely to be common to decision making in the construction industry, is the argument that some carbon footprints can be limited to CO_2 if the emissions of the other gases in the Kyoto basket are known to be negligible. However, in these instances it is essential to justify and document any and all exclusions, and any 'rules of thumb' used in producing a footprint (Mackenzie, 2011).

Double counting is when the same emissions are included more than once in the same footprint and may occur for a range of reasons, for example where two or more reporting bodies within an organisation share responsibility for an emissions source. Conversely, there is the risk that attempts to avoid double counting may lead to an emissions source being omitted completely if both reporting bodies ascribe the source to the other. To avoid this it is useful to map all reporting bodies and emissions sources as part of the development of a footprint, and justify which emissions sources should be attributed to which organisational body, or if (and how) an attribution should be shared. This is particularly problematic where organisations are engaged in partnerships or sub-contracting agreements, and the GHG Protocols offer detailed guidance on how to manage accounting under these circumstances (WRI, 2004).

One argument against the value of carbon accounting that arises from this complexity is that the knowledge gaps and the uncertainties inherent in different approaches and tools mean that none will produce a precise and complete figure for CO_2e. So as the earth doesn't care about humans getting the figures right, our efforts would be better invested in simply doing everything we can to reduce those emissions. However, in the real world the resources we have at our disposal to do so are limited, and the justifications for using them must be based on the most robust and scientifically sound evidence available. Furthermore the rationale behind the development and deployment of that evidence must be consistent in order to avoid undermining public trust and provoking accusations of 'green-washing', which may be overcome through greater transparency (Davis, 2011).

Enabling transparency highlights one conflict amongst those aiming to improve the overall quality and usefulness of carbon footprints. At present those seeking to develop or commission a footprint are faced with the question of whether to use one of the many off-the-peg accounting tools or to opt for developing a bespoke approach, either in-house or using external expertise. The costs involved in developing and maintaining footprinting tools mean that many contain components, for example databases of conversion factors, that are locked from access by the end users. This means that those without this access may see little more than a 'black box', i.e. data goes in one end and results come out the other without the user having full knowledge of what has gone on inside. In most cases such black boxes will be retained for their value as intellectual property (IP), especially in the case of authoritative third party providers who gain revenue from licensing their tools. However, this can lead to accusations over lack of transparency, even if the tools have been subjected to independent inspection by appropriate and impartial peer reviewers. Retaining and capitalising on IP is a particularly tricky problem owing to the intricacies of the patent and copywriting laws, as exemplified by the number of lawsuits between software companies. Another reason for locking tools is the risk that non-expert groups, particularly climate change deniers, may inspect them and use any perceived errors to further their own agendas.

Arguably, the benefits of enabling greater transparency outweigh those of retaining opacity. The actual value of the IP contained in a tool may not be as high as its developers may think, especially if part of that value is in the form of revenue gained through providing training and support materials, i.e. the 'real' value may be in the credibility of the authors and not in the tool itself, which could simply be replicated by a sufficiently knowledgeable competitor. Furthermore, greater transparency can be achieved whilst retaining some locked components if these are opened up for review by appropriate experts. However, the overriding benefit of transparency, or at least being as transparent as possible, is ensuring accountability and gaining stakeholder and public confidence, both in the results of carbon footprints and in the decisions that are based upon them.

Finally, the benefit of ensuring accuracy is that any errors in a footprint are systematic rather than random. For example, it is much easier to correct a conversion factor if it is used throughout a footprint, which in the case of a bespoke spreadsheet based assessment could be controlled for by having all conversion factors listed in a page of cells which is linked to every instance of their use, i.e. it should be much easier to alter one value that updates a set of pages than to check them all individually. Another benefit of this approach arises where ownership or use of a tool or inventory changes between people and/or organisations over time, for example through staff turnover or mergers between organisations. Put simply, there is little value in repeating a foot-printing exercise if the results are not directly comparable over time. Therefore in order to enable greater accuracy it is essential to document every component of a footprinting tool and ensure any revisions are recorded, dated and justified (Baker, 2011).

11.2 C, CO_2, CO_2e and GHG emissions factors

The amount of carbon emissions calculated as part of an assessment is usually expressed in one of three ways:

1 *Carbon (C)*: This is the amount of carbon atoms only and is increasingly rare. To convert to CO_2 multiply this by $\frac{44}{12}$ (or 3.67).
2 *Carbon dioxide (CO_2)*: This is the amount of carbon dioxide only, and is useful for accounting exercises in which the contribution of other GHG emissions is known to be negligible. It is still widely in use.
3 *Carbon dioxide equivalent (CO_2e)*: This is the amount of carbon dioxide plus the relative amounts of the other gases in the Kyoto basket, which are weighted by their global warming potential. This is the gold standard for carbon accounting and should be calculated wherever feasible. It is now the most common figure for expressing GHG emissions.

When developing or using a carbon footprinting tool, or comparing their outputs, it is essential to understand and account for which of these units are in use, and ensure that their usage is clear and consistent throughout. It is also essential to understand where and if any proxies or approximations have been used, for example if GHGs other than CO_2 have been measured directly or based on the assumption of being relative to CO_2.

CO_2e is calculated by taking the amount of each GHG in the Kyoto basket and multiplying by a conversion factor that accounts for its impact on climate change relative to that of one unit of CO_2, modelled over a period of 100 years. The relative impact of one unit of CO_2 is 1. For example, the global warming potential for methane over 100 years is 23. This means that emissions of 1 million metric tons of methane are equivalent to emissions of 23 million metric tons of carbon dioxide. The current conversion factors are given in Table 2.1.

Conversion factors for calculating CO_2 or CO_2e directly from common sources such as energy generation and transportation are now widely available from sources including the UK Department for Energy and Climate Change (DECC, 2010), the Australian Department for Climate Change (DCC, 2009), the US Environmental Protection Agency (EPA, 2004) and the IPCC (1996).

When using factors such as these it is important to ensure that they are the latest and most relevant available. For example, the conversion factors for emissions from electricity production vary according to different national energy mixes, and factors for transportation vary nationally according to the average fuel efficiencies of common modes of transport. Whenever used they should be fully documented and referenced, along with details of any subsequent revisions. If a particular set of conversion factors is not available it may be possible to derive the factors from first principles, or it may be sufficient to use factors from elsewhere, depending on the needs of the exercise and the resources available. It may also be useful to provide figures for energy and resource consumption in base units (GWh, Btu and so on) alongside emissions data, or to dis-aggregate CO_2e into the individual gases. The latter is particularly useful where higher volumes of one or more GHGs are known to be emitted, for example methane emissions from farming livestock. Another common way in which emissions can be represented is per unit of population (CO_2e per capita), which is commonly used when comparing the relative differences in emissions between countries.

An alternative way of expressing environmental impact is in global hectares (gha). This gives the area of land that is needed to provide the energy and resources necessary to sustain, for example, an individual, a city or even the entire population of the planet. assessments of this are known as 'ecological footprints' and are popular for communicating environmental impacts to non-specialist audiences.

In addition, many government bodies and other organisations offer free online carbon calculators, but most of these are unlikely to be sufficient for the needs of professional carbon accountants. An exception to this is the Stockholm Environment Institute's highly sophisticated Resource and Energy Analysis Programme (REAP) developed for the UK (see Chapter 10 for details), which is also available in a more basic form (REAP Petite) suitable for smaller organisations (SEI, 2011).

11.3 Baselining

Baselining is the process of establishing the historical emissions for a given entity, for example an organisation or a city, normally calculated for the year prior to engaging in a voluntary or mandatory emissions reduction strategy. Producing a robust and accurate baseline is essential, as subsequent emissions accounting needs to be comparable in order to demonstrate any emissions reductions, and any inaccuracies may invalidate or undermine the level of progress achieved. In the case of cap and trade carbon markets, such as the EU's Emissions Trading Scheme (ETS), credits are allocated according to the individual baselines of participants, and therefore underestimating can have costly repercussions. A standard baseline can be produced by following five basic steps:

- *Step 1:* Decide and justify source boundaries.
- *Step 2:* Identify CO_2 emission sources.
- *Step 3:* Decide base year(s).
- *Step 4:* Compile data for base year(s).
- *Step 5:* Estimate emissions and quality-assure data (Carbon Trust, 2011).

11.4 Scoping emissions

Scoping emissions is the process of deciding and justifying which sources of emissions should be included in a baseline and subsequent emissions accounting; therefore from the outset of a project producing a robust set of scopes is essential to ensuring its validity long after the initial baselining work. Any significant changes in scope may ultimately result in the need for a new baseline.

A standard scoping process first identifies and categorises emissions as direct and indirect, which are defined by the GHG Protocols group as follows:

- *Direct GHG emissions* are emissions from sources that are owned or controlled by the reporting entity.
- *Indirect GHG emissions* are emissions that are a consequence of the activities of the reporting entity, but occur at sources owned or controlled by another entity.

The GHG Protocol further categorises direct and indirect emissions into three broad scopes:

- *Scope 1:* direct GHG emissions emitted at the point of combustion of fuels;
- *Scope 2:* indirect GHG emissions from consumption of purchased electricity, heat or steam (i.e. direct GHG emissions from the production of electricity, heat or steam);
- *Scope 3:* indirect emissions such as the extraction and production of purchased materials and fuels, transport related activities in vehicles not owned or controlled by the reporting entity, electricity related activities (e.g. transmission and distribution – T&D – losses) not covered in Scope 2, outsourced activities, waste disposal and so on.

Emissions data for direct CO_2 emissions from biologically sequestered carbon (for example, CO_2 from burning biomass or biofuels) are reported separately from these scopes.

These scopes provide a common framework for carbon accounting that is widely in use around the world, but exact inclusions, exclusions and allocations may vary and are likely to be revised in the future.

A key issue to consider here is the level of control an organisation has over each emissions source, which is essential for determining the boundaries for a carbon accounting exercise.

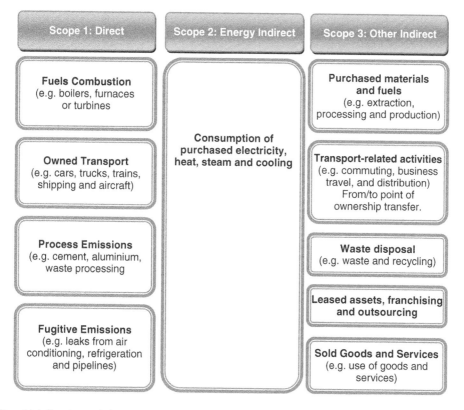

Figure 11.1 Scoping emissions

Source: Based on WRI, 2004; DECC, 2010.

11.5 Setting boundaries

Determining and justifying boundaries is a fundamental stage in emissions accounting. Depending on the project it will be necessary to determine boundaries according to the level of control over an emissions source (see 11.4 above) and/or the lifecycle boundaries for a building or product. For the former, where sources are shared between two or more entities the GHG Protocols set out a framework that can be applied based on either equity share or financial control (see Table 11.1). More commonly, those working in the built environment will need to justify the lifecycle boundaries for carbon accounting. These can also be used for, and should be consistent with, the boundaries used for other forms of lifecycle accounting (LCA).

Figure 11.2 shows a typical framework for setting lifecycle boundaries for a building, and how they relate to those for products, which include building components. Figures for the energy and emissions embodied in many building materials are available in the Inventory of Carbon and Energy (ICE) developed at the University of Bath (Hammond and Jones, 2011).

Table 11.1 Emissions accounting guidelines for shared sources

Accounting category	Accounting for GHG emissions according to the GHG Protocols corporate standard	
	Based on equity share	*Based on financial control*
Group companies or subsidiaries	Equity share of GHG emissions	100% of GHG emissions
Associated or affiliated companies	Equity share of GHG emissions	0% of GHG emissions
Non-incorporated joint ventures, partnerships or operations where partners have joint financial control	Equity share of GHG emissions	Equity share of GHG emissions
Fixed asset investments	0% of GHG emissions	0% of GHG emissions
Franchises	Equity share of GHG emissions	100% of GHG emissions
Leased assets	See detailed guidance in WRI (2006).	

Source: Based on CDP, 2011; WRI, 2004.

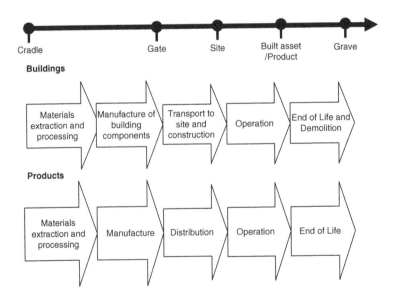

Figure 11.2 Lifecycle boundaries for carbon accounting

Using only the ICE data, data for types and modes of transportation, and the appropriate conversion factors it is possible to produce a robust estimate of the energy and emissions embodied in many buildings and products. Note that most, but not all, of the figures given in the ICE database are for 'cradle to gate', and in any case calculating emissions from transportation separately using real data should produce more accurate results.

Operational emissions are generally modelled using one or more models or tools (see Chapter 10), but ideally this should be supported by measured data after the building has been occupied. It is important to note the distinctions between embodied and operational emissions and modelled and measured emissions, especially in cases where buildings are claimed to be net 'low' or 'zero' carbon, as many published figures refer to modelled operational energy only.

In addition, the operational emissions from a building would also include the emissions embodied in any maintenance to the building over its lifespan, although these are not yet commonly included in assessments.

Finally, for buildings it is difficult to account for the emissions embodied in the end of life phase owing to the uncertainty of not knowing how and when the building will be demolished and how the waste material will be processed. For this reason many experts, such as the ICE authors, advise scoping this phase out of lifecycle emissions assessments for buildings.

As part of scoping and determining boundaries it is likely that uncertainties will arise that may necessitate the use of assumptions or proxies, or even the exclusion of one or more components of the emissions inventory. These problems may arise for a variety of reasons, including:

*   difficulty in gathering data;
*   metering or measurement issues;
*   data management constraints;
*   incomplete information for the period;
*   structural changes (mergers, acquisitions or divestments);
*   outsourcing and/or insourcing of activities;
*   unreliable information.

As with all aspects of carbon accounting, whenever these problems occur it is essential to document and justify how they have been addressed, particularly in the case of any exclusion that may relate to conditions placed on new buildings or products. However, developing and maintaining a full inventory of materials and processes may aid compliance with other requirements, for example legislation around the use of potentially hazardous materials.

Another option for 'reducing' the emissions from buildings, products and organisations that may also need to be factored into emissions assessments is the use of carbon offsetting.

11.6 Carbon offsetting

Carbon offsetting is the practice by which individuals or organisations invest in reducing emissions that are then counted against their own carbon footprint. For those participating in carbon trading schemes these investments are converted into credits. Carbon offsetting is highly controversial for a number of reasons, but chiefly because those who engage in it have effectively decided that it is a cheaper and/or easier option than reducing their own emissions. Those who argue against the validity of carbon offsetting frequently point to the numbers of offsetting projects based in the developing world (and are therefore in conflict with the 'polluter pays' principle) and/or those that involve tree planting (because of the areas of land required and uncertainties over the carbon embodied in timber – for more on the latter see 8.1.1, 'Carbon in building materials'). Nonetheless, carbon offsetting is now widely in use both voluntarily and as an 'allowable solution' to reducing emissions in carbon trading and reduction schemes.

Carbon credits can take the form of allowances (or rights to pollute) or offsets, which are project based GHG emissions reductions programmes (Bayon, Hawn and Hamilton, 2009). Offset projects can create carbon credits by reducing any of the six GHG gases listed in the Kyoto Protocol – carbon dioxide, methane, nitrous oxide, sulphur hexafluoride, hydrofluorocarbons and perfluorocarbons – and they can do so by emissions reduction of these six GHGs, by destruction of existing GHGs or by reduction of GHGs through sequestration projects (Liverman, 2009). Figure 11.3 illustrates these project categories.

Offset programmes can use any of these projects or their combinations to generate carbon credits that are tradable under a growing number of schemes, ranging from international markets such as the EU ETS to community level carbon reduction projects. Carbon credits are usually generated in one of two forms: cap and trade systems and baseline and credit systems.

11.6.1 Cap and trade

This system places an overall limit on the level of emissions allowed to a group of entities. Regulatory bodies then distribute these allowances, or rights or permits of emissions, within these entities (Crossley, 2008). Authorisations to emit in the form of emission allowances are allocated to affected sources, and the total number of allowances cannot exceed the cap or the overall limit set (SEI, 2008). Permit holders who are able to reduce their emissions internally through efficiency or other emission reduction mechanisms can sell their unused allowances at the market price (Bayon *et al.*, 2009). The underlying market making mechanism of this system is based on sound economic principles backed by the fundamentals of effective demand and supply. As there is a limited supply of allowances which is predetermined by country level policy negotiations and binding by regulation, these allowances can thus not be created or destroyed, but can only be traded (Kollmuss *et al.*, 2010). Such systems are typically regulated by government and require mandatory compliance. The only main challenge to them is the arbitrariness and the discretion used to determine the appropriate level of setting the cap, which should be both effective and capable of delivering the desired reductions.

11.6.2 Baseline and credit system

In a baseline and credit system there is no limit to the number of carbon offsets that can be produced (Crossley, 2008). Credits are generated whenever emissions are offset or reduced through

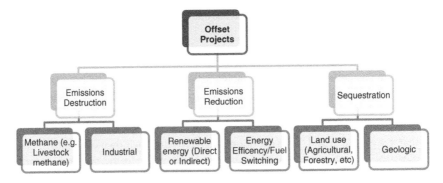

Figure 11.3 Common offset project categories

Source: Derived from, Bayon *et al.* 2009; Ademola, 2011.

the implementation of an offset project. It is notable, however, that projects under the cap and trade system cannot be counted towards baseline and credit system reductions (MacGill, Outhred and Nolles, 2004). Credits generated in this manner can be used by purchasers to comply with a regulatory emission target to 'offset' their emissions or to be 'carbon neutral' by reducing the amount of emissions they produced to net zero (Crossley, 2008; Kollmuss *et al.*, 2010). This system uses a project based baseline scenario to calculate the level of emissions that would occur had the offset project not been implemented.

Table 11.2 presents the distinguishing features of cap and trade and baseline and credit systems.

For those wishing to assess the credibility of carbon offset projects, the following is a non-exhaustive list of requirements that provide a basic framework to evaluate them against:

- *Accounting:* The amount of emissions reduction claimed for a project should be measured directly (for example, from energy data) and/or based on a robust carbon accounting assessment.
- *Additionality:* There must be proof that the reduction in emissions would not have happened without a carbon offset project, and the project is not a result of some legal obligation or compliance against legally binding targets.
- *Permanence:* The amount of emissions reduced should be reliable and permanent in the sense that it achieves the goals it is committing to, and impermanent projects like forestry projects that are at a risk of disease or fire must be addressed and, if need be, compensated with replacement of credits by the project developer or offset seller.
- *Avoiding double counting:* The seller must ensure that each carbon unit is removed from the tradable market to avoid double counting or double selling.
- *Timing:* Ideally, the sale of carbon credits must always be after the project has been completed and the emissions reduction has taken place.
- *Leakages:* The project must demonstrate that there is no leakage such that the reduction in one area should not lead to an increase in emissions elsewhere. Leakage occurs when carbon

Table 11.2 Distinguishing features of cap and trade and baseline and credit systems

Features	Cap and trade	Baseline and credit
Exchanged commodity	Allowance.	Carbon credits.
Quantity available	Determined by overall cap.	Generated by each new project.
Market dynamic	Buyers and sellers have competing and mutually balanced interests in allowances trades.	Buyers and sellers both have an interest in maximizing the offsets generated by the project.
Sources covered	Usually high emitters such as energy and energy intensive industries.	As defined by each standard. Not limited to just high emitting sectors.
Independent third party	Minor role in verifying emissions inventories.	Fundamental role in verifying the credibility of the counterfactual baseline and thus the authenticity 'additionality' of the claimed emissions reductions.
Emissions impact of trade	Neutral, as is ensured by zero sum nature of allowance trades.	Neutral, providing projects are additional. Otherwise, net increase in emissions. Possible decrease in emissions in the voluntary market.

Source: Ademola, 2011.

saving measures employed by a project lead to increased emissions elsewhere. Should this be unavoidable these should be measured and removed from sale.

- *Independent evaluation:* The project should be independently evaluated with respect to all of the above, including repeat evaluations for those projects where emissions reductions are subject to any risks or uncertainties (Ademola, 2011; Bayon *et al.*, 2009; DECC, 2009; Kollmuss *et al.*, 2010; Peters, 2008; Valatin, 2009).

11.7 Reporting

Reporting is the task of pulling everything together: hence the importance of accurately recording all stages of a carbon assessment. The diversity of sectors and organisations now needing to or wishing to produce their own assessments means that there is no standard reporting format, but some examples are available, such as PwC (2011).

When drafting a report it is advisable to consider:

- *Scale:* The larger and more complex the assessment, the more detail that may be needed in the report.
- *Importance:* Bear in mind the purpose of the report and its target audience.
- *Stakeholders:* Consider both who should be consulted as part of compiling the report, and how it may be interpreted by external bodies, interest groups and the wider public.
- *Potential for reductions:* If further reductions could be made or will be made in future these should be noted, along with the justification for any prior aims and objectives that have not been met.
- *Contractual agreements:* The content of the report should comply with the reporting requirements of any schemes it is being used for, and proof of compliance with any requirements placed on sub-contractors (e.g. offsetting projects) should be verified and documented.

References

All websites were last accessed on 30 November 2011.

Ademola, A. 2011. Carbon offsetting by UK retailers. M.Sc. dissertation, School of the Built and Natural Environment, Glasgow Caledonian University, Glasgow.

Baker, K.J. 2011. *Opening Black Boxes: Enabling Greater Transparency and Accuracy.* Edinburgh and Glasgow: ICARB and Glasgow Caledonian University. Available at: icarb.org/2011/05/24/the-need-for-transparency-keith-baker/.

Bayon, R., Hawn, A. and Hamilton, K. 2009. *Voluntary Carbon Markets*, 2nd edn. London: Earthscan.

Carbon Trust. 2011. *Emissions Baseline Guide.* London: Carbon Trust. Available at: http://www.lowcarbon-cities.co.uk/cms/assets/Toolkit-Documents/2-Assess-City-State-of-Play/Emissions-Baseline-Guide.pdf?phpMyAdmin=b2269d0b74f0653ad75eab0d458e983d.

CDP (Carbon Disclosure Project). 2011. *Accounting, Boundaries and Scopes.* London: CDP. Available at: https://www.cdproject.net/en-US/Respond/Documents/Webinars/2011/accounting-boundaries-and-scopes.pdf.

Crossley, D.J. 2008. Tradeable energy efficiency certificates in Australia. *Energy Efficiency*, **1**, pp. 267–281.

Davis, G. 2011. *Carbon Footprinting: Setting the Boundaries.* Edinburgh: Edinburgh Centre for Carbon Management/Econometrica. Available at: http://icarb.org/2011/02/15/boundaries-gary-davis/.

DCC (Australian Department for Climate Change). 2009. *National Greenhouse Accounts (NGA) Factors.* Canberra: DCC. Available at: http://www.climatechange.gov.au/~/media/publications/greenhouse-gas/national-greenhouse-factors-june-2009-pdf.ashx.

DECC (Department for Energy and Climate Change). 2009. *Guidance on Carbon Neutrality.* London: DECC. Available at: http://www.decc.gov.uk/assets/decc/What%20we%20do/A%20low%20carbon%20UK/carbonneutrality/1_20090930090921_e_@@_carbonneutralityguidance.pdf.

DECC. 2010. *2010 Guidelines to Defra/DECC's GHG Conversion Factors for Company Reporting.* London: DECC. Available at: http://archive.defra.gov.uk/environment/business/reporting/pdf/101006-guidelines-ghg-conversion-factors.pdf.

EPA (US Environmental Protection Agency). 2004. *Unit Conversions, Emissions Factors, and Other Reference Data.* Washington, DC: EPA. Available at: http://www.epa.gov/cpd/pdf/brochure.pdf.

Hammond, G. and Jones, C. 2011. *Inventory of Carbon and Energy.* Bath: University of Bath. Available at: http://www.bath.ac.uk/mech-eng/sert/embodied/.

IPCC. 1996. *Revised 1996 IPCC Guidelines for National Greenhouse Gas Inventories: Workbook.* Geneva: IPCC. Available at: http://www.ipcc-nggip.iges.or.jp/public/gl/guidelin/ch1wb1.pdf.

Kollmuss, A., Lazarus, M., Lee, C., LeFranc, M. and Polycarp, C. 2010. *The Handbook of Carbon Offset Programs.* London: Earthscan.

Liverman, D. 2009. Carbon offsets, the CDM and sustainable development. Available at: http://ele.arizona.edu/files/ELEliverman4-17-09B.pdf.

MacGill, I., Outhred, H. and Nolles, K. 2004. Some design lessons from market-based greenhouse gas regulation in the restructured Australian electricity industry. *Energy Policy,* **34**, pp. 11–25.

Mackenzie, C. 2011. Comparing corporate performance on climate change – what metrics? Carbon Benchmarking Project, University of Edinburgh. Available at: http://icarb.org/2011/02/14/comparing-corporate-performance-craig-mackenzie-draft/.

Peters, G. 2008. Reassessing carbon leakage. Paper presented at the Eleventh Annual Conference on Global Economic Analysis: Future of Global Economy, Helsinki, June.

PwC (PricewaterhouseCoopers). 2011. *Carbon Reporting Report.* London: PwC. Available at: http://www.pwc.co.uk/eng/publications/carbon_reporting.html.

SEI (Stockholm Environment Institute). 2008. *Carbon Offset Markets 101*, presentation. Stockholm: SEI. Available at: http://www.epa.gov/cmop/docs/cmm_conference_oct08/14_kollmus.pdf.

SEI. 2011. *REAP: Resource and Energy Analysis Programme.* Stockholm: SEI. Available at: http://sei-international.org/reap.

Valatin, G. 2009. *Carbon Additionality: A Review.* Edinburgh: Forestry Commission. Available at: http://www.forestry.gov.uk/fr/INFD-7WUEAN.

WRI (World Resources Institute). 2004. *The Greenhouse Gas Protocol: A Corporate Accounting and Reporting Standard*, rev. edn. Washington, DC: WRI.

WRI. 2006. *The Greenhouse Gas Protocol*, Appendix F: Categorising GHG emissions from leased assets. Washington, DC: WRI. Available at: http://www.ghgprotocol.org/calculation-tools/all-tools.

12 Carbon and GHG accounting for organisations

Accounting for the carbon and other GHGs attributable to an organisation (or company) can be a highly complex task that requires a series of steps to identify, measure and attribute emissions and ultimately compare them both over time and between comparable organisations. The 'gold standard' method for developing an organisational carbon footprint is the GHG Protocol Corporate Accounting and Reporting Standard (WRI, 2004), which despite its name can also be used for public and third sector organisations, and therefore the terms 'organisation' and 'company' are used interchangeably here and throughout this text, unless specified otherwise.

The focus of this text on the built environment means that some readers may have little need for engaging in organisational carbon and GHG accounting. However, it is useful to have some knowledge of the processes, as accounting for the emissions from built assets may feed into wider organisational footprints. Therefore this chapter provides a summary guide to how to go about developing a carbon footprint for an organisation, but for guidance on particularly complex issues such as attributing emissions across a multi-entity organisation or partnership it is advisable to consult the more specific guidance provided in the supplementary documents to the GHG Protocol Standard.

12.1 Scoping and setting organisational boundaries

The most critical task in developing a robust organisational carbon accounting system is that of scoping and setting boundaries. See Figure 11.1 for a (current) definition of the three scopes used in accounting.

When scoping and setting boundaries it is essential to understand what the development of a carbon or GHG footprint report is intended to achieve. As a minimum this will be the following:

- identifying and defining what the organisation will measure, report, manage, reduce and possibly offset over time, i.e. what emissions the organisation is directly responsible for, and what indirect emissions (beyond Scope 2) it has some level of control over;
- producing a robust, accurate, transparent and complete assessment of the emissions attributable to an organisation based on a consistent methodology that allows year on year comparisons to be made;
- a footprint report that is able to identify and manage risks that could lead to increased emissions, and able to identify opportunities for reducing emissions from organisational activities;
- a footprint that also allows comparisons to be drawn over time and with other organisations.

A footprint report may also be intended to, or required to, achieve any of the following and more:

- demonstrate and document compliance with mandatory or voluntary emissions reduction systems;
- enable participation in carbon or GHG markets;
- provide proof of the impact of new actions taken to reduce emissions;
- communicate the results to the public and non-specialist audiences.

The following case provides a very basic example of scoping emissions for a small company, in this case an independent restaurant. However, the questions and accounting tasks become increasingly more complex the larger the organisation, the more sites it operates, the more it is organisationally subdivided, and the more it engages in partnerships with other organisations. Therefore in order to deal with this complexity larger organisations can be broken down into accounting units, usually described by legal, financial or geographic boundaries. This not only simplifies the accounting process but also enables greater transparency by allowing for the attributing of emissions across and between organisations, and can be valuable for identifying and avoiding potential cases of double counting. Further difficulties, and risks for double counting, are likely to arise where a company or organisation produces components that are part of a supply chain for other products. Attributing emissions between organisations working in partnerships is a particularly complex task for which the guidance is still evolving. This is covered in Chapter 11 (see 11.5, 'Setting boundaries'), but readers are strongly advised to consult the supplementary documents available from the GHG Protocols group for the latest advice on best practice.

12.1.1 Note on Scope 3 emissions

The Scope 3 ('other indirect') emissions category uses the same definition as prescribed by the earlier work of the GHG Protocols group, and is the same as that currently used by many compliance systems, for example the UK's CRC Energy Efficiency Scheme (covered in the case study at the end of the chapter). However, the classification of emissions under Scope 3 is contested, and new work by the GHG Protocols group and other leading organisations is clarifying and redefining emissions covered by this scope, and some would argue that the scope should be removed and all indirect emissions grouped under Scope 2. Recently the GHG Protocols group have published two new guidance documents that are intended to help clarify and resolve the issues raised by the definition of Scope 3 and the problems of accounting within product chains and for the use of ICT (WRI, 2011, 2012). These should be referred to for developing robust carbon and GHG accounting systems for more complex organisations.

12.1.2 Example of a simple organisational scoping exercise

For a small organisation such as an SME, the boundaries may be as simple as direct and indirect emissions plus most or all of the Scope 3 emissions. For example, consider a small independent restaurant. The restaurant is unlikely to use on-site combustion (unless any non-mains supply fuels are used for cooking), so Scope 1 emissions may be limited to emissions from vehicles owned by the company. It may also leak some fugitive emissions from refrigeration, but these are likely to be negligible and difficult to measure with any degree of accuracy, and so can be scoped out. All Scope 2 emissions can simply be assessed from recording consumption of electricity and natural gas and applying the appropriate conversion factors, some of which could then be offset.

Scope 3 presents more of a problem, as here the restaurant would need to determine which emissions categories it has insufficient control over that may justify exclusion. Clearly any business travel would be included, but what about emissions from employees commuting to work? If all employees walk, cycle or use public transport then it might be fair to say the company has done all it can to reduce these emissions, and as any further reductions would be beyond its control these can be scoped out. However, unless the restaurant has a specific policy requiring employees not to commute by car it might still be worthwhile to report these as zero but still allow for the use of private cars to be accounted for in the future. It could also be assumed that all food is consumed on or near the site and/or that the restaurant has no control over how customers take their food away, and therefore the transport and distribution of purchased goods beyond point of sale can be scoped out.

However, for a more complete footprint, emission from the transport of goods to the restaurant, along with its own embodied emissions, would need to be scoped in. If the restaurant has no plans to expand through franchising or engage in outsourcing then these can be excluded too, but if it does expand this way they would need to be scoped back in later – hence the importance of documenting and justifying any changes to boundaries that may occur over time. Ideally any emissions attributable to the use of products on site, such as cleaning products, would need to be included, but it is unlikely that a small retailer would have access to sufficiently accurate data for this to be useful, and so these could be justifiably scoped out.

Finally, the restaurant will produce waste that will itself be responsible for emissions. Although it is unlikely that a small retailer would be required to account for these emissions, doing so may identify new opportunities for reducing them. For example, the restaurant may dispose of waste through a standard municipal collection network that ultimately ends in a landfill, but emissions reductions could be claimed if some or all of this waste is diverted to recycling, composting or waste to energy streams.

12.2 Protocols, standards and systems

There is a plethora of protocols, standards and systems with which organisations may have to or choose to comply, and the number is still growing. To complicate matters further, organisations operating around the world may have to comply with different combinations of these according to the regions or countries they operate in. Although some organisations are working towards a more unified accounting framework, new research and development in carbon and GHG accounting is still likely to lead to the development of new protocols, standards and systems and revisions of existing ones. Table 12.1 gives a selection of these, including the most important international ones. However, this is a non-exhaustive list and limited to those published in English. It should also be noted that many of these are not restricted or specific to carbon and GHG accounting, and may also include assessing factors such as sustainability and impacts on biodiversity.

Table 12.1 Selected examples of protocols, standards and systems

Country or region	Organisation	Standard(s)
Global	IPCC	Guidelines for National GHG Inventories.
	United Nations	Clean Development Mechanism.
	WRI/WBCSD	GHG Protocol.
		GHG Project Accounting Standard.
		GHG Scope 3 Accounting Standard.
		GHG Protocol Product Life Cycle Accounting and Reporting Standard ICT Sector Guidance.
	ISO	ISO 14064-1: Specification with guidance at the organisation level for quantification and reporting of greenhouse gas emissions and removals.
		14064-2: Specification with guidance at the project level for quantification, monitoring and reporting of greenhouse gas emission reductions or removal enhancements.
		ISO 16064-3: Specification with guidance for the validation and verification of greenhouse gas assertions.
	Climate Group, International Emissions Trading Organisation, World Economic Forum and WBCSD	Verified Carbon Standard (formerly the Voluntary Carbon Standard).
	Carbon Disclosure Project	Carbon Disclosure Project – global climate change reporting system.
	Plan Vivo	Plan Vivo.
	Global Reporting Initiative	G3 Guidelines.
	Climate, Community and Biodiversity Alliance	CCB Standards.
	Gold Standard Foundation	Gold Standard.
North America	US Environmental Protection Agency	Greenhouse Gas Reporting Program.
	Climate Registry	Climate Registry.
	Western Climate Initiative	Western Climate Initiative Cap and Trade Program.
	Center for Resource Solutions, California	Green-e Climate.
	Climate Action Reserve, California	Climate Action Reserve.
	CarbonNeutral	The CarbonNeutral Protocol.
South America	Ecologica Institute, Brazil	Social Carbon.
EU	EU	European Emissions Trading System (EU-ETS).
		EU technical guide for the calculation of the environmental footprint of companies.
		EU harmonised methodology for the calculation of the environmental footprint of products.
UK	Department for Energy and Climate Change	CRC Energy Efficiency Scheme (formerly Carbon Reduction Commitment).
	Department for the Environment, Food and Rural Affairs	Guidelines for Company Reporting on GHG Emissions.
	Carbon Trust	Carbon Trust Standard.
		PAS 2050: Assessing the Life Cycle Greenhouse Gas Emissions of Goods and Services.
		PAS 2060: Specification for the Demonstration of Carbon Neutrality.
	Forestry Commission	Woodland Carbon Code.
France	Bilan Carbone	Bilan Carbone.

Sources: Table compiled with help from Gary Davis, Director, Ecometrica.

Note: As of December 2011.

12.3 Tools for organisational accounting

The rapid proliferation of organisational protocols, standards and systems has fuelled an equally rapid proliferation in the number of tools that can be used to comply with them. Globally the most widely used protocols are those of the GHG Protocols group. However, protocols only set a framework with which tools should comply. Meeting some standards, such as the Carbon Trust's publicly available specification (PAS) standards, requires following a set accounting procedure which must be conducted and/or certified by a registered assessor, but others are more flexible in the methods and results they accept.

This need for expertise has led to many organisations and companies forming or expanding to meet the demand. Some of these are consultancy based, whereas others sell specific software, commonly linked to an online database of factors that may or may not be disclosed to users. Perhaps the most well-known and highly regarded of these is the suite of tools developed and licensed by the Resources and Energy Analysis Project (REAP) at the Stockholm Environment Institute, University of York, UK (SEI, 2008) (see also Chapters 10 and 11 for this and other tools). Many organisations and companies choose to develop their expertise and systems in-house, either by recruiting specialist staff or by buying in training and specialist consultancy. This means that it is common for an organisation to develop its own carbon and GHG accounting system from scratch using only professional guidance and a spreadsheet program.

Regardless of which option is chosen, the costs of producing a footprint for a large organisation have inevitably led to some loss of transparency as intellectual property rights are asserted over new tools. How to enable greater transparency whilst still incentivising the development of new and improved accounting tools and methods is a key issue for the future of carbon and GHG accounting.

12.4 Cross-sector and sector-specific tools

Regardless of how they work in practice, most organisational GHG accounting tools come in one of two forms, cross-sector and sector-specific, and most organisations need a combination of these to complete a full emissions assessment. Cross-sector tools cover common emissions sources such as the consumption of electricity and fossil fuels by buildings and vehicles. Tools tailored for individual sectors are usually needed to account for sources such as process and fugitive emissions. In some cases, for example an entirely office based organisation, it may be sufficient to use only a cross-sector tool, but deciding which tool or tools to use should always require scoping the emissions of an organisation first.

For an example of an organisation needing both types of tools, consider an aluminium production facility. This would need a cross-sector tool for stationary combustion (for purchased electricity, on-site generation of energy, and so on) and for mobile combustion (for transportation of materials by train, on-site vehicles, employee business travel, and so on). It would also need a sector-specific tool to account accurately for emissions of PFCs specific to aluminium production (CDP, 2011).

12.5 Engaging staff in organisational carbon and GHG accounting

A potential barrier to those developing organisational carbon and GHG footprints is that some staff, particularly senior staff who may be new and potentially resistant to carbon accounting, may not appreciate the level of engagement necessary to produce an accurate and robust footprint. Increasingly, larger organisations are creating specific departments to manage their emissions reduction and sustainability obligations. Whilst these are undoubtedly important for ensuring that adequate time and resources can be directed at meeting these obligations, they risk creating 'silos' from which it may be difficult to disperse knowledge and expertise more widely. Furthermore, in a large organisation it is highly unlikely that a specialist staff team will be able to capture sufficient knowledge of organisational emissions, and opportunities for emissions reduction, without significant engagement with other departments. Such departments may include, but not be limited to, the following:

* business operations;
* environment;
* facilities management;
* finance;
* human resources;
* legal;
* procurement;
* sales and marketing.

If in doubt about the need to engage with any departments or key staff it should always be assumed they should be included (DECC, 2010). An important measure in enabling wider engagement in organisational carbon management is those mandatory emissions reporting and reduction compliance systems that include financial penalties and rewards based on performance. The bottom line here is that legislating for emissions reductions will inevitably require the use of large financial sticks until sufficient awareness and knowledge of carbon and GHG accounting become normalised at all levels of large organisations.

Case study: An organisational compliance scheme: the UK's CRC Energy Efficiency Scheme

The UK's CRC Energy Efficiency Scheme (formerly the Carbon Reduction Commitment and still commonly termed the CRC) provides a useful case study of a national organisational carbon reduction compliance system. As large organisations vary widely in their nature and operations, providing a case study of one would be of limited value here. However, the CRC provides a useful example of a compliance and trading scheme for emissions reduction, as it is highly prescriptive in detailing how to account for and attribute emissions in large and complex organisations whose composition is likely to change over time. This is necessary to support the development and operation of a robust national emissions trading market, and the basic process is common to other national and international market led schemes.

The CRC entered into force in 2008 and aims to deliver emissions savings of at least 4Mt CO_2 per year by 2020 using a mandatory auction based cap and trade scheme covering large business and public sector organisations, which account for around 10 per cent of the UK's emissions. The scheme is mandatory for all organisations that had at least one half-hourly meter (HHM) settled on the half-hourly market as of 2008, and whose half-hourly metered electricity consumption was greater than 6,000MWh/year for 2008.

Once entered into the scheme an organisation must account for *all* emissions from its energy consumption directly attributable to its activities in the UK, excluding only transport, accommodation, energy supplied to third parties, and emissions already covered by climate change agreements and the EU ETS.

In common with other emissions trading systems, participation in the CRC requires an organisation to follow five key steps:

1   Forecast future emissions.
2   Purchase allowances (credits – equivalent to 1 tonne of CO_2) to cover those emissions, based on the results of a footprinting exercise.
3   Measure, monitor and manage emissions so as not to exceed the forecast.
4   If necessary, purchase additional allowances if actual emissions are higher than forecast.
5   Report organisational emissions and fully document evidence of compliance at the end of each accounting period.

Once initial allowances have been purchased through a closed auction system, participants are able to trade allowances on the Secondary Market, which is also open to traders from organisations not covered by the CRC, but as allowances are cancelled at the end of each reporting period they cannot be 'banked' for future use or sale. The opening up of the Secondary Market and the cancelling of allowances raise the possibility of environmentally minded philanthropists buying up allowances in order to retire them permanently from the market, thereby driving up the cost of carbon; and the cap system means that in theory the market could be exhausted before all organisations have purchased enough additional allowances to cover their emissions. Therefore in order to prevent this happening the scheme includes a 'safety valve' mechanism where organisations can request the release of additional allowances, at a price linked to the cost of carbon on the EU ETS, but one which must be higher than the CRC sale price. In addition, the performance of organisations participating in the CRC is ranked in a publicly available league table that is intended to stimulate further emissions reduction through competition (DECC, 2010).

At the end of each three-year accounting and trading period, participating organisations are ranked in a league table, with those at the bottom subject to financial penalties for poor performance. Strict penalties can be enforced for non-compliance, ranging from small fines for late submissions up to large fines and prison sentences for deliberate non-compliance or falsification of results. In its original form the CRC was also intended to recycle the penalties for poor performers as bonuses for those at the top of the table, giving an added incentive for improving performance. However, this added benefit was soon dropped and has led to accusations that the CRC is now merely a 'carbon tax'.

The first CRC league table was published at the end of the first phase of the scheme in 2011, and the increased, or otherwise, performance of participating organisations in future league tables will be an important indicator of the effectiveness of market based approaches to reducing emissions from large organisations.

Note that at the time of going to press revisions to the CRC were undergoing consultation.

References

All websites were last accessed on 30 November 2011.

CDP (Carbon Disclosure Project). 2011. *Accounting, Boundaries and Scopes*. London: CDP. Available at: https://www.cdproject.net/en-US/Respond/Documents/Webinars/2011/accounting-boundaries-and-scopes.pdf.

DECC (Department for Energy and Climate Change). 2010. *The CRC Energy Efficiency Scheme: User Guide*, 6 April update. London: DECC. Available at: www.decc.gov.uk/assets/decc/what%20we%20do/a%20low%20carbon%20uk/crc/1_20100406154137_e_@@_21934crcpdfawv9.pdf.

SEI (Stockholm Environment Institute). 2008. *REAP*. York: Stockholm Environment Institute, University of York. Available at: http://www.resource-accounting.org.uk/reap.

WRI (World Resources Institute). 2004. *The Greenhouse Gas Protocol: A Corporate Accounting and Reporting Standard*, rev. edn. Washington, DC: World Resources Institute/World Business Council for Sustainable Development. Available at: http://www.ghgprotocol.org/files/ghgp/public/ghg-protocol-revised.pdf.

WRI. 2011. *Corporate Value Chain (Scope 3) Accounting and Reporting Standard*. Washington, DC: World Resources Institute/World Business Council for Sustainable Development. Available at: http://www.ghgprotocol.org/files/ghgp/Corporate%20Value%20Chain%20%28Scope%203%29%20Accounting%20and%20Reporting%20Standard.pdf.

WRI. 2012. *GHG Protocol Product Life Cycle Accounting and Reporting Standard ICT Sector Guidance*. Washington, DC: World Resources Institute/World Business Council for Sustainable Development. Available at: http://www.ghgprotocol.org/.

Index